The way customers find businesses has changed — and most small business owners haven't been told how.

If you're a small business owner, coach, consultant, or independent professional, you've likely felt it: endless effort creating content that doesn't seem to add up, a cycle of showing up and starting over, and the sense that you're visible—but not clearly understood. You may have even tried AI tools, only to get results that feel generic or off-target.

This isn't a motivation problem. It's a clarity problem.

SEEN: How Small Businesses Get Seen, Found, and Chosen in AI Search offers a different way to think about visibility—one that fits how discovery works now. Customers now ask AI tools for answers — and those systems interpret, compare, and narrow options before a person ever clicks a link. Visibility is no longer only about posting more or chasing trends. It's about being understandable.

Businesses that are clear, consistent, and easy to describe are easier for customers to trust—and easier for systems to categorize and carry forward.

In this practical, research-informed book, S. R. Prater shows you how to:

Understand how AI-powered search actually works — and why clarity now determines whether your business gets recommended.

Clarify what you do so customers (and AI systems) recognize it quickly.

Reduce mixed messages that quietly undermine trust and visibility.

Build recognition that compounds over time instead of resetting every week.

Create Human-First, AI-Readable content without sounding robotic.

You won't find hacks, heavy tech explanations, or pressure to "be everywhere." You'll find clear frameworks, real examples, and simple exercises designed for busy owners who want their marketing to feel lighter—and work better.

If AI can't read you, it can't recommend you. This book shows you how to become readable—on your terms.

SEEN

How Small Businesses Get Seen, Found, and Chosen in AI Search

S.R. PRATER

SunajMedia, LLC

SunajMedia, LLC

Published by Sunaj Media, Omaha, NE

Library of Congress Control Number: 2026903115

ISBN: 979-8-9945359-0-5

Front Cover, Spine and Back Cover illustrated by B.M. Hardin

Printed in the United States of America.

This publication is designed to provide accurate and authoritative information regarding the subject matter covered. It is intended for educational and informational purposes only.

The author and publisher are not engaged in rendering legal, financial, investment, medical, or other regulated professional services through this publication. The concepts, frameworks, and examples presented are general in nature and are not a substitute for personalized professional advice.

While the author may offer professional coaching, consulting, or advisory services separately, this book does not establish a coach–client, consultant–client, or other professional relationship. Readers should consult qualified professionals regarding their specific circumstances when appropriate.

AI technologies, platforms, and search behaviors evolve continuously. The information in this book reflects conditions at the time of publication and may not reflect future developments.

The author and publisher make no representations or warranties regarding the accuracy or completeness of the content and disclaim any implied warranties of merchantability or fitness for a particular purpose. Neither the author nor the publisher shall be liable for any damages arising from the use of this material.

This book was written by a human author, with AI used as a supportive tool — not a replacement for thinking, judgment, or lived experience. AI tools were used during the research, drafting, and editing of this book. These tools assisted with organizing ideas, refining language, and testing clarity — the same principles the book teaches. All frameworks, interpretations, conclusions, and recommendations reflect the author's professional experience, research synthesis, and original analysis.

https://www.seensmallbusiness.com

Contents

To My Mom, Lynette

"Will they make it so we can get it"

Foreword

The systems people use to find what they need are changing faster than most people have time to notice.

That's not a criticism. It's just true. If you're running a business, raising a family, going to school, or doing all three—you're not spending your evenings reading about how AI search works. You're doing your life. Technology shifts anyway.

This has happened before. When I was a student we learned the Dewey Decimal system because it was required. But once I was in the library, searching for books, I liked how it worked. You'd look up one book, and the system would show you what was next to it. You could see what was related. You could wander into something you didn't know you needed. The system didn't just help you find a book—it helped you discover.

Then search engines came along. Then social media. Then algorithms. Each time, the rules for being found changed, and each time, most people had to figure it out while everything else in their lives kept moving.

We're in another one of those shifts right now. AI tools are changing how people find businesses, compare options, and make decisions. These systems don't just show a list of links—they summarize, compare, and recommend. And they do it before a person ever visits your website or reads your reviews.

What makes this shift different is where the decision-making happens. It's moving upstream—into systems most people never see and don't fully understand. That's not because people aren't smart enough. It's because these changes weren't announced with clear instructions.

Meanwhile, the places that used to help people make sense of complex systems are disappearing—or being restructured in ways that change who they naturally serve. Public libraries are one of the last truly open spaces where anyone could walk in and learn how things work. A few years ago, a city I know well demolished a beloved downtown library to make way for commerce. The city built a smaller replacement in the area, and a new state- of- the art library in another part of the city. The library didn't vanish. But the one embedded in people's daily routines—where a teenager could duck in from the cold to wait for the bus, where a young parent could bring their children to learn something for free —that version was gone. The replacement serves a different moment, a different part of town, a different need.

This book comes from that same sense—the belief that when something important changes, someone should make it understandable for the people who need it.

I wrote SEEN because small business owners deserve a clear explanation of what's happening, not a technical manual designed for enterprise teams with dedicated staff and six-figure budgets. The technology behind AI discovery is complex, but the principles that make a business easier to find are not. They're about clarity, consistency, and making sure your message can be understood by both the people you serve and the systems now shaping how those people decide.

You don't need to become an AI expert. You don't need to learn to code. You don't need to chase every platform change. You need your business to be easy to understand —and you need a calm, clear guide to show you how.

This book has been accepted into the Library of Congress—which, for someone who understands what public libraries make possible, feels like the kind of full circle that's hard to put into words. I hope it finds the people it was written for.

— *S.R. Prater*

How to Use This Book

Let me tell you the purpose of this book in plain language. It's twofold.

First: most people are still not truly AI-proficient. A lot of the workforce is using AI casually—if they're using it at all—and many employees don't feel prepared or confident with it. That's according to publicly available research.

If that's you, you're not behind—you're normal. And if you're a small business owner, you're probably trying to run a business and keep up with a new technology shift at the same time.

So this book gives you more AI understanding than the average person who "uses AI sometimes." It's not a prompting book, and it's not a technical manual. But it will give you the fundamentals you need to make sense of what's changing—so you can make better business decisions without feeling lost.

Second: this book is here to help you transition your visibility and marketing into the new discovery environment. Because the way customers find businesses is changing fast. Search isn't just links. Systems are summarizing, comparing, and recommending. And there is a real risk of being left behind—not because your business isn't good, but because it isn't being interpreted clearly by the systems helping customers decide.

Quick Start: If You Only Have 15 Minutes

If you want the fastest way to make this book useful immediately, go to the Bonus Checklist first. *The checklist is most useful after you understand Chapters 1-3, but if you need immediate triage, start there and return here.*

It will help you spot mixed messages and contradictions across the places customers and systems check—your website, profiles, listings, and reviews. When those are aligned, your business becomes easier to recognize, trust, and recommend.

Use the Parts Like a Menu

You don't need to read this in order. Start with the part that matches what you're dealing with right now:

Part I — Understanding the Shift (Ch. 1–3) Start here if you want to understand why visibility has changed and learn the foundational frameworks for clarity: the Topic Trio, Question Bank, and the power of one idea at a time.

Part II — Building Clarity (Ch. 4–7) Start here if people don't understand what you do quickly, your message feels heavy, or you keep rewriting the same descriptions.

Part III — From Clarity to Stability (Ch. 8–12) Start here if you want to understand what AI systems trust, how to optimize without fragility, and how to build visibility that compounds over time.

The Goal

This book isn't about doing more marketing. It's about making your business easier for customers—and systems—to understand.

That's how your visibility becomes steadier.

That's how trust becomes faster.

That's how the right people find you and choose you without constant output.

Why This Book Matters Now

We are entering a new frontier in how people find businesses.

For most of modern history, discovery was physical. Customers found businesses through location, signage, word of mouth, and local reputation. Visibility depended on proximity and presence.

The internet changed that. Discovery moved online, and search engines became the primary way people found answers to their problems. Businesses learned to think in terms of websites, keywords, and rankings. Being found meant appearing in a list of links and hoping someone clicked through.

Then social platforms changed discovery again. Recommendations became social. Visibility became tied to activity, posting frequency, and engagement. Businesses were encouraged to show up constantly—to be seen, liked, and shared in real time.

Each shift didn't replace the one before it. It layered on top of it.

What's happening now is another transition.

This moment is part of a broader shift often described as the Fourth Industrial Revolution—where technology doesn't simply add new tools, but reshapes the systems underneath everyday life: how people learn, how they shop, how options are compared, and how decisions are made.

AI is now part of that system.

Tools like ChatGPT, Perplexity, and Google AI Overviews don't work like traditional search engines. They don't just display a list of links. They generate answers. They summarize. They compare options. They recommend selections—often before a person ever visits a website or profile.

This changes what it means to be findable.

Visibility is no longer just about showing up. It's about being understood well enough for a system to describe your services and products accurately. Businesses aren't only competing for attention; they're being interpreted, categorized, and compared by systems designed to reduce choice and simplify decisions.

That shift is subtle, happening now, and it's foundational. And it's why clarity—how clearly a business can be explained, confirmed, and summarized—now determines whether it gets carried forward in the discovery process. In industry research, this shift is often called Generative Engine Optimization, or GEO. This book uses simpler language — AI Visibility — but the principles are the same: making your business clear enough for AI systems to understand, categorize, and recommend.

Let me show you why this matters with a real example that changed how I think about discovery. I first became aware of this shift on an ordinary summer day.

I opened an AI platform to do some early holiday shopping and asked a simple question: Where can I find the best hot chocolate? I expected a short list of familiar brands and locations. Instead, the response arrived like a guided recommendation.

It didn't just answer. It clarified what "best" meant.

The system asked follow-up questions about preferences—dark chocolate or milk, sweetness level, how well it dissolves, texture, richness, and whether I cared about details like fair trade, organic ingredients, or artisan production. It referenced articles on the topic, then returned specific recommendations with product details—prices, images, and links to retailers where I could buy them.

In a few minutes, I learned about styles and quality tiers I didn't even know existed—hot chocolate made from real chocolate shavings, blends built around flakes, even mixes designed for a thicker, smoother finish. Each response felt increasingly curated and increasingly clear.

What stood out wasn't just the convenience. It was the introduction to brands I had never encountered before—presented in a way that felt confident, structured, and easy to understand.

So I tried it again with other products I was shopping for. That's when I realized: this wasn't just a faster way to search. It was a different way to shop.

And it changed my perspective.

I stopped thinking like a consumer and started thinking like a business owner.

How did these brands get categorized so clearly?

Why did some show up in the results while others weren't mentioned at all?

What made the system confident enough to recommend certain products as "best"?

Those questions sent me into deeper research, long periods of synthesis, and sustained systems thinking. What I found was strikingly consistent across industries and sources:

In AI-driven discovery, being everywhere matters less than being understandable.

These systems rely on patterns. They scan for clear descriptions, stable categories, consistent language, and information that aligns across public sources. When a business is easy to interpret and confirm, it becomes easier to recommend.

That is why I wrote this book.

SEEN gives you a clear, durable way to communicate what you do so you can be found, confirmed, and chosen—by people first, and by the systems now shaping discovery. You don't need to become technical. You don't need to chase every trend. You need clarity that holds steady. This is the new visibility frontier.

Who This Book Is For

This book is for small business owners, solopreneurs, coaches, and creators who want a clear, simplified understanding of how modern discovery works—without needing a technical background. If you've felt like visibility has become harder to predict, or like your message keeps needing "one more rewrite," *SEEN* helps you make your business easier to understand, easier to confirm, and easier to recommend—using plain language and a calm, practical approach.

Who This Book Is Not For

This book is not for people looking for hacks, shortcuts, or a fast platform game. It also isn't a technical SEO manual or a guide to building AI tools. If you want a sustainable method—one that reduces confusion, stabilizes your message, and helps the right people find you without constant output—you're in the right place.

About the Author

After years working in data systems and teaching professionals how to adapt to emerging technology, I became fascinated by a new question: ***How do AI systems decide which businesses to recommend?*** I researched how these systems interpret information, studied the patterns that create visibility, and realized most small businesses were struggling not because their work wasn't good—but because their message wasn't structured for how AI actually reads. This book is the result of that research, written for business owners who don't have time for technical jargon or enterprise playbooks.

S.R. Prater is a Certified AI Coach, Certified Acquisition Advisor, consultant, and educator. With a background in data systems and training leadership—including teaching soft skills to students in an emerging technology program as an adjunct instructor—she helps professionals and business owners adapt to technological change with clarity and focus.

Additional resources related to this book are available at seensmallbusiness.com.

A Note on Examples

Throughout this book, you'll find examples of businesses applying these principles—including bakeries, coaches, fitness studios, and service providers. These examples are modeled on real patterns observed across multiple clients and industries, supported by research into how AI systems interpret content. They illustrate how clarity principles work in practice. Individual results will vary based on industry, location, and how consistently the methods are applied.

An Ethical Approach to AI

This book follows an ethical, human-first approach to AI use. The author holds an **ICF-accredited AI coaching certification** and adheres to professional coaching standards that emphasize transparency, responsibility, and human agency.

In practice, this means:

AI is treated as an assistive tool—not an authority

Human judgment remains central to decision-making

No personal or client data was used to train or personalize AI outputs

The intent is empowerment, not dependency

The approach modeled in this book reflects how AI can be used **responsibly, ethically, and with respect for human autonomy**—especially in business, coaching, and creative work.

Did You Know: AI Summarizes Before It Recommends
Before AI recommends your business, it creates an internal summary. If your content mixes multiple ideas, that summary becomes blurred — and a blurred summary rarely gets carried forward into a recommendation. The businesses that get recommended are the ones AI can summarize in one clear sentence.

Chapter One

Why Visibility Changed

A New Kind of Visibility

Visibility looks different than it once did.

For years, small businesses relied on familiar pathways to be discovered—word of mouth, social media, traditional search, and consistent posting. While these channels still matter, the way people find information has shifted quietly and quickly. A growing number of customers now turn to AI tools to explore options, compare services, and make decisions.

This is a new kind of visibility: **AI Visibility**—the ability for your business to be seen, found, and chosen within AI-driven systems. You may hear this called — Generative Engine Optimization or GEO. It's the industry term for what happens when AI systems summarize and recommend instead of just ranking links. This book focuses on the practical side: how to be understood by those systems.

Business owners may feel this change before they can name it:

- Content feels heavier

- Effort feels less effective

- Visibility seems harder to maintain—even when you're doing more than ever

It isn't because they're doing something wrong—it's because the environment that determines visibility has changed.

The Core Change

Traditional web search (especially platforms like Google and Bing) works like this:

You search → you get links → you decide what to click

AI-powered search works differently:

You ask a question.→ The system reads many sources. →It gives you an answer.

That answer may cite sources, summarize them, or recommend next steps — but the decision-making happens before you ever click.

AI systems like ChatGPT and Claude can function as discovery tools and can operate similarly to AI-powered search interfaces like traditional web search.

That's the shift. Discovery used to start with a list of options and leave the choosing to you. Now, systems are interpreting, comparing, and narrowing before you see anything. What reaches you has already been filtered. This process is called **upstream interpretation** — the system evaluates and summarizes information before a person ever sees it. In traditional search, people made choices from a list. In AI-powered search, the system makes interpretive choices first.

This means visibility is no longer just about showing up. It's about being understood clearly enough to make it through that filter.

How GEO Relates To SEO

If you've spent time on SEO—optimizing your website for search engines—that work still matters. GEO doesn't replace SEO. It builds on top of it.

SEO helps your website appear in search results. GEO helps AI systems understand, summarize, and recommend your business when people ask questions instead of typing keywords.

Think of it this way:

SEO focuses on ranking—getting your page to appear in a list of links.

GEO focuses on clarity—making sure AI can accurately describe what you do when it generates an answer.

Both matter. But as more people use AI tools to research, compare, and decide, clarity becomes as important as ranking.

You don't need to abandon your SEO efforts. You need to make sure the content you've already created is clear enough for AI systems to interpret and carry forward.

How AI Search Actually Works

According to a recent article published on , 65% of customers now ask AI for recommendations, the way they used to ask a trusted friend. They open ChatGPT, Perplexity, or Google's AI Overviews and type questions like "best running shoes for beginners" or "local bakery with gluten-free options"—often without realizing they've left traditional search behind.

Here's what changed: AI doesn't reward effort or volume. It rewards clarity.

Think of it like this: when someone asks you what you do at a party, a clear answer sticks. A scattered one doesn't. AI works the same way—except it's reading your website, your social posts, your listings, and your reviews all at once, looking for one consistent story.

What AI notices (when it's scanning your content):

- Do you talk about the same topics consistently?

- Does your language stay recognizable, or does it shift every week?

- Can it tell what you offer and who it's for—quickly?

When your message stays clear and focused, AI can describe you confidently. When it shifts too often—mixing unrelated ideas, changing your wording, or trying to be everything to everyone—AI hesitates. Stanford University research confirms this: AI systems often misinterpret unclear or loosely structured writing—not because the AI is flawed, but because the content lacks the clarity needed for accurate interpretation.

Your content doesn't disappear. It just becomes harder to summarize, harder to categorize, and harder to recommend.

This is what we call a **visibility gap**: the space between what you're trying to say and what AI systems can actually understand.

How AI Shows Business Recommendations

What is the best automotive repair place in Kansas City, Missouri that works on foreign sports cars?

Autobahn Automotive
Foreign & Exotic Car Specialists
9876 Broadway Blvd, Kansas City, MO
📞 (816) 555-5678
**4.7 "Great Porsche and BMW repairs"

Precision Import Service
Luxury & Performance Car Repair
4567 Wornall Rd, Kansas City, MO
📞(816) 555-7890
**4.9 "Top-notch work on Ferraris and Audis!"

EuroTech KC
Specializing in European Sports Car
1234 Main St, Kansas City, MO
📞 (816) 555-1234
**4.8 "Expert service, highly recommended"

The AI system matched the search request to businesses whose descriptions clearly stated: automotive repair Kansas City foreign vehicles Businesses that do not consistently express those signals are filtered out — even if they offer the service. AI recommendations depend on structured clarity, not reputation alone.

What AI Search Results Look Like

When customers open AI tools and ask:

- *"What is the best automotive repair place in Kansas City, that work on foreign sports cars?"*

- *"Where can I buy handmade candles?"*

- *"What business coaches help with leadership transitions?"*

They don't see a list of 10 blue links.

Instead, AI generates a synthesized response that looks more like a personalized recommendation (like what you see in the picture above):

Here are three local automotive mechanics known for foreign car repair work:

- *EuroTech KC — specializes in European sports cars*

- *Autobahn Automotive — foreign and exotic car specialists*

- *Precision Import Service — luxury and performance car repair*

If your business is clear, consistent, and easy to categorize, AI tools can include you in answers like this—even if you don't rank on Google, don't have a large audience, or don't post every day.

Industry data shows that customers who click links recommended by AI often convert at higher rates than traditional search traffic. **AI tools can only recommend what they can recognize and categorize accurately.**

How AI Actually Recommends Businesses

AI tools typically recommend businesses for two main reasons:

Recommendation Type #1 — Trusted Across the Web

AI often recommends a business because it appears steady across trusted sources: review platforms (like Google or Yelp), business directories, reputable websites, press mentions, public records, brand managed information and consistent business listings across platforms.

When these sources line up, AI gets a clear picture of who the business is, that it can be trusted, and that customers can depend on it.

Recommendation Type #2 — Clear, Understandable, Easy to Categorize

AI also recommends businesses simply because it can clearly understand what they do. This is the most controllable factor for small businesses.

AI surfaces your business when:

- Your language is clear

- Your topics are consistent

- Your services are easy to understand

- Your website and posts align

- You express one idea at a time

- Your content is structured for easy categorization

When AI encounters your business, it's checking three things::

1. What does this business do?

2. Who is it for?

3. When should this business be recommended?

Even a very small business can outperform a larger competitor in AI Visibility when its message is clear, consistent, and easy for AI systems to categorize. *(Yext Research. AI Doesn't Rank, 2025)*

This shift—from traditional search to AI-driven recommendations—isn't speculation. It's already happening.

The Data Behind The Shift

Multiple research sources confirm this change is real and accelerating. Constant Contact's Small Business Now:2024 report shows small businesses increasingly struggle with keeping their content discoverable and adapting to AI tools.

These shifts happened quietly, without public guidelines, and small businesses feel the effects because they don't have teams dedicated to monitoring or adapting to constant changes.

Larger companies figured this out early. They standardized their messaging, kept their language consistent, and made their content easy to scan and summarize. Industry reports confirm it: clear, well-organized content that repeats across sources gets cited far more often in AI-generated answers.

The New Invisibility

When AI tools can't clearly understand a message, they may not include that business in their answers—even when the business offers strong and valuable services.

You may notice signs of this when:

- Content tasks don't feel "done"

- Updates feel constant

- Ideas don't fully land the way you intended

- Visibility seems tied to constant posting

- Your message feels scattered or hard to summarize

These are not signs of failing. They're signals that your message may need clearer structure and more consistent cues.

Here's the good news for small businesses. You have an advantage large companies don't. You can align your message faster, with fewer layers and less bureaucracy.

Why Small Businesses Need a Different Approach

Most AI visibility advice comes from enterprise marketing play-books—strategies designed for teams, budgets, and high-volume content engines.

Enterprise strategies assume:

- Publishing across every channel

- Constant content production

- Teams managing SEO, video, podcasts, PR

- Long timelines for data to be ingested into AI systems

Small businesses need something different:

- Clarity instead of volume

- Consistent signals instead of constant publishing

- Simple structures AI can easily interpret

- A sustainable rhythm instead of a content engine

Your visibility should not depend on being everywhere—it should depend on being clear.

Chapter 1 Key Takeaways:

- AI systems can function as discovery tools when connected to live data.

- AI can only recommend what it can clearly recognize and categorize

- Small businesses can align their message faster than large companies

- The challenge is not effort—it's clarity and consistency

Chapter 2 shows you what to do about it—starting with one simple clarity tool.

Chapter Two

What You Do With That Information

The Topic Trio and Customer Question Bank

The changes described in Chapter 1 aren't calling for more effort, more posting, or more platforms. They're asking for something simpler: **focus**.

AI Visibility grows from clarity. Clarity grows from consistency. And consistency becomes easier when you narrow your message to a small set of related ideas that reinforce one another over time.

This chapter introduces two practical tools that make your expertise easier to see, easier to find, and easier to choose: the **Topic Trio** and the **Question Bank**.

Common Patterns That Blur Your Message

Before we build clarity, let's identify what works against it. These aren't mistakes in effort—they're simply patterns that can blur your message over time:

These patterns in this table show why focus outperforms volume.

Common Pattern	How It Shows Up	Why It Makes Visibility Harder	What to Do Instead
Scattered focus	Posting about many different themes before previous one form a pattern	AI systems struggle to group your work into one clear category	Choose 2-3 related topics and stay with them long enough to build recognition
Multiple Ideas in one message	A single post or paragraph tries to explain several concepts at once	Mixed ideas are harder to label, summarize and remember	Give each idea it's own post or paragraph
Inconsistent language	Using different terms to describe the same service, or explaining your work with a new angle every time	Different wording can look like different services, and without repeated phrasing, recognition builds more slowly	Pick one clear phrase for each service and reuse it — start from your core sentence
Trying to appeal to everyone	Broad messaging aimed at 'anyone; who needs help	Broad positioning makes it harder to connect your work to a specific need	Define who is the clearest starting point that your work fits best

Common Patterns Matrix

If you offer multiple services, each one can have its own clear explanation. The goal is to avoid blending them in a single message where none land clearly.

How AI Identifies Patterns

AI tools identify your message by detecting four main patterns:

Repeated Ideas – Topics you revisit regularly help establish your expertise. Each repetition reinforces what you know and what you offer, allowing people and AI tools to connect you with the subjects you care about most.

Consistent Themes – Related topics that appear regularly help form a clearer picture of what you do.When your themes stay connected, both people and AI systems begin to associate you with those specific areas of expertise.

Recognizable Language – Similar phrasing creates stability and makes it easier for systems to interpret what you mean. Over time, familiar wording becomes a cue that reinforces who you are and what you offer.

Clear Focus – A small set of well-defined topics helps AI connect your content to specific questions and needs.

When your focus stays consistent, it becomes easier for AI systems to understand where your expertise fits and when your business should be recommended.

Throughout this book, you'll see the same principles applied again and again: writing for people first, structuring information so it's easy to interpret, and repeating what matters instead of reinventing it.

You don't need to memorize this approach yet—you'll recognize it as we go. Small shifts in communication can create large shifts in clarity.

Tool #1: The Topic Trio: A Simple Clarity Framework

Your first clarity tool is called the **Topic Trio**. It focuses your entire message into three core elements:

1. **What you help with** (the problem or goal)

2. **Who you help** (your audience)

3. **The method or perspective you bring** (your unique approach)

These three elements become the anchors of your message. When they stay steady, your content becomes easier for people and platforms to follow and categorize.

Example (how the 3 questions guide a Home + About page) Imagine you're a web designer helping a client who teaches beginners how to create digital products. Your client wants their website to do one main job: help people immediately understand what they do—so customers (and AI systems) can categorize them correctly.

So as you build the **Homepage** and **About page**, you keep three questions in front of you—not as marketing theory, but as a writing filter.

The three (3) questions you're answering on the page

1. **What do they help with?** Digital product creation

2. **Who do they help?** Beginners and freelancers

3. **How do they help? (method)** Simple, step-by-step teaching

These three answers become the "spine" of the page. Everything else supports them.

What That Looks Like In Actual Website Copy

HOMEPAGE (TOP SECTION)

Headline: Create your first digital product—without getting overwhelmed.

Subhead: I help beginners and freelancers turn ideas into simple digital products they can sell online, using a step-by-step process.

ABOUT PAGE (FIRST PARAGRAPH)

Hi, I'm [Name]. I teach **beginners and freelancers** how to create and sell their first **digital product**. My approach is **simple and step-by-step**—so you always know what to do next, even if you're starting from scratch.

Notice what's consistent:

- The same three elements (digital products, beginners/freelancers, step-by-step) appear in *both* the headline and the About page

- The language stays recognizable—no synonym switching

- A customer reading either page would describe this business the same way

This is what clarity looks like in practice.

Because when a website repeatedly answers what / who / how in plain language, it becomes easier for customers to trust—and easier for AI systems to summarize accurately.

Create Your Topic Trio

*This **3-minute** exercise anchors everything that follows. Without it, Chapters 3-12 will feel harder than they need to be.*

Fill in these three sentences:

- **I help people with**: _____

- **I help people who are**: _____

- **My core perspective or method is**: _____

Now combine them: *"I help [who] with [what] using [method]."*

This becomes your clarity anchor. Keep it handy—you'll use it in every chapter that follows."

In Chapter 1, you learned that **how people find businesses** is now question-driven. Search used to be keyword-driven. Now, AI tools surface content that answers the kinds of questions people ask. This is sometimes called **answer-first search**. Instead of returning a list of links for you to browse, the system generates a direct response — a summary, a comparison, or a recommendation. The user receives an answer as the starting point, not a page of options.

For small businesses, this is an advantage. You already answer questions about your business every day through conversations, emails, social posts, and service work.

Why Questions Matter for AI

AI tools find content that:

- responds directly to a question

- uses clear, simple phrasing

- stays focused on one idea

- matches topics you talk about repeatedly

When you write around actual customer questions, your content becomes:

- easier for people to understand

- easier for platforms to categorize

- easier for AI to recommend

When your content mirrors real customer questions:

-AI can recognize it as a helpful answer and recommend it.

That's where the **Question Bank** comes in.

Customer Questions Are Your Answers

Sarah's Small Bakery – One Sentence Change Everything

Sarah runs a small bakery in Portland. For years, her Google Business Profile said: "We create beautiful, artisan baked goods for special occasions and everyday treats."
After completing the Topic Trio exercise, she rewrote it to: **"I help families celebrate moments by baking fresh, made-to-order desserts."**

Within 60 days:
- Her Google Business Profile views increased 40%
- AI tools (ChatGPT, Perplexity) started recommending her when people searched "custom birthday cakes Portland" or "fresh desserts near me"
- She got 3 inquiries that specifically mentioned "I found you through an AI search"

The Three Types of Customer Questions

To make this practical, here are the three categories of questions customers typically ask—and each one fuels a different type of AI-readable content.

A. Questions customers actually ask you

These come from:

- conversations

- emails

- DMs

- client sessions

- sales calls

- comments on posts

They reflect immediate, real-world needs.

Examples:

- "Do you offer this service for beginners?"

- "How does your pricing work?"

- "What's the difference between A and B?"

These questions build trust and clarity.

B. Questions customers don't know to ask but need answered

These are based on your expertise.

Examples:

- "What should someone look for before hiring a ___?"

- "What mistakes should beginners avoid?"

- "What do most people misunderstand about ___?"

These questions position you as a guide.

C. Questions people ask AI tools about your category

These shape your discoverability.

Examples:

- "Best fitness trainer near me for beginners"

- "How do I start a clothing boutique?"

- "How do I write clearer content for my business?"

Questions like these help AI link your business to real user needs. **That's why you're about to build a Customer Question Bank.** This is one of the simplest, most powerful tools for AI-readable visibility.

Tool #2: Build Your Customer Question Bank

(10-15 Minutes)

Step 1 — Questions Your Customers Actually Ask

Think about the questions you already answer in conversations, emails, DMs, consultations, or comments. These questions come directly from customers and reflect real curiosity or confusion.

Write three questions your customers actually ask you:

- _____

- _____

- _____

Step 2 — Questions Your Customers Need Answered (But Don't Know to Ask)

These questions come from your experience working with customers—the things they should know, but rarely think to ask.

Write three questions your customers need answered:

- _____

- _____

- _____

Step 3 — Questions People Might Ask AI Tools

These are the questions people type into AI tools when seeking clarity. Review the example below, to understand how to create your questions for step 3.

Example: Women's Clothing Boutique

People might ask AI questions like:

- "What clothing styles are best for petite women?"

- "Where can I find affordable, high-quality everyday outfits for women?"

- "How do I build a capsule wardrobe for work and weekends?"

- "What should I look for to make sure clothes fit well when shopping online?"

These questions reveal curiosity, uncertainty, and a desire for guidance by the shopper—and make excellent sources of content topics.

If you are the business answering these questions, your answers help AI understand:

- what category your business belongs to

- who your content is meant for

- what problems you help people solve

Here are few helpful starter prompts to begin:

- "Best ___ for ___"

- "How to start ___"

- "How to fix ___"

- "What's the simplest way to ___?"

Write three (3) questions people might type into AI:

_____

_____

_____

You will use this Customer Question bank to:

- Write clearer posts

- Create helpful content

- Structure simple stories

- Write your first AI-readable article

- Repurpose content across platforms

Each question becomes an anchor point that keeps your content focused and easy for people—and AI tools—to understand.

When clear topics, repeated themes, consistent language, one idea at a time, and simple structure all come together—AI matches your message to real user questions more reliably. This is how some small businesses show up in AI search results.

This is why your Question Bank becomes a foundational tool for visibility.

Chapter 2 Key Takeaways:

- Clarity comes from focus, not volume

- The Topic Trio builds your message anchor

- The Question Bank turns your expertise into AI-readable content

- People remember connected ideas more easily than scattered ones

- These two tools make every chapter after this easier

Chapter Three

One Idea At A Time

Helps People and AI Understand What You're Saying

Clarity isn't just about *what* you say—it's about how you structure it. This chapter shows why one idea at a time works better than cramming everything into one message. Imagine you run a small service business—like an eyeglass shop.

Customers reach out for all kinds of reasons:

- a refund for a defective pair of glasses

- a question about their prescription

- help adjusting their frames

- pricing or insurance questions

- an update on when their order will be ready

If you answered all of those requests the same way, customers would get frustrated fast. Not because you don't care—but because their specific issue isn't being handled.

When you know what someone needs—a refund, an adjustment, or an order update—helping them is simple. Writing works the same way.

When you separate ideas and handle them one at a time, your message becomes easier to understand—for people and for the AI systems trying to figure out what you do.

AI Clarity Self-Check *(This is a rapid self-inventory of beliefs and clarity around AI visibility — no expertise required.*

1.When I hear "AI visibility" I feel:
☐ Curious
☐ Overwhelmed
☐ Skeptical
☐ Neutral

2. My current understanding of AI in my work is:
☐ I don't think it applies to me
☐ I know it matters, but I don't know how
☐ I use AI tools occasionally or regularly
☐ I can explain AI visibility to someone else

3. When describing my business or what I do:
☐ I use different language on different platforms
☐ I'm not sure how to simplify it
☐ I can describe it clearly in one sentence
☐ People often misunderstand it
☐ People repeat it back accurately

4. My current state of affairs with AI:
☐ I've been trying to "keep up"
☐ I've been avoiding learning more
☐ I've been quietly experimenting
☐ I've been pretending I understand more than I do
☐ I'm ready to get clear instead of reactive.

Notice what you checked. Clarity begins with recognition

Content Edges

Clear Edge   ### Crossed Edge

This post is about
choosing the right
running shoes for
beginners

This post is about running
shoes, marathon training tips,
and staying motivated

Why edges matter On the left, the message stays inside one clear boundary. One topic. One purpose. One signal. On the right, three different ideas compete for attention. When a post crosses edges, readers hesitate — and AI systems do too. Clarity strengthens recognition. Mixed signals weaken it. AI categorizes one idea at a time.

How One Idea at a Time Works

People are pretty good at following along, even when writing jumps around.

We fill in the gaps. We connect ideas. We usually get what someone means, even if it's a little messy.

AI doesn't do that.

AI looks for clear signals. It tries to figure out:

- what this is about

- who it's meant for

- what question it's answering

When a paragraph tries to say several things at once, those signals get mixed together.

The words are still there—but the meaning becomes harder to pin down.

That's not because AI is broken.

It's because it needs clearer structure than people do.

When you write one idea at a time, you remove the confusion.

Your message becomes easier to understand, easier to summarize, and easier to surface when someone is looking for help.

Before-and-After Example

Here's a sentence written with good intentions—but too much packed into it:

Mixed version:

"I help small business owners clarify their messaging, build confidence in their content, and create systems that reduce overwhelm while improving visibility."

A person can follow this.

AI has a harder time deciding what this is really about.

Now look at the same idea, broken apart:

Clear version:

"I help small business owners clarify their messaging."

"Clear messaging reduces overwhelm."

"When messages are easier to understand, visibility improves."

Nothing new was added.

Nothing important was removed.

Each sentence now handles one idea—which makes the message easier to understand, easier to reuse, and easier for AI to recognize.

Your First Clarity Shift

Here's the shift this chapter is asking you to make.

You're not trying to sound better.

You're trying to be clearer.

When you slow down and separate your ideas:

- people understand you faster

- your message sticks

- writing feels easier, not heavier

Clarity doesn't come from saying more.

It shows up the moment each idea has its own space.

Practice: The One Idea Check

After you write a sentence or paragraph, pause and ask:

"What am I really trying to say here?"

If the answer includes more than one main topic, split it up.

That's it.

You don't need to do this all the time.

Use it when your message feels heavy, confusing, or hard to explain.

Content Edges: Why One Idea at a Time Works

When you write one idea at a time, you're creating what this book calls a **Content Edge—a clear boundary around one topic.**

A **content edge** answers:

- What is this about?

- What is this NOT about?

When content stays inside a clear edge:

- People follow it more easily

- AI systems can categorize it accurately

- Repetition strengthens recognition instead of creating confusion

When content crosses edges (mixing multiple topics):

- Readers get confused

- AI systems hesitate

- Your message scatters instead of compounds

You don't need many edges. You need a few clear ones, repeated calmly.

Example:

- **Clear edge:** "This post is about choosing the right running shoes for beginners"

- **Crossed edge:** "This post is about running shoes, marathon training tips, and staying motivated"

The first has one boundary. The second tries to hold three different ideas—and none of them land clearly.

In the next chapters, you'll use content edges to make your message easier to repeat without confusing your audience.

The One-Sentence Test

After you write a paragraph, try to summarize it in one calm sentence.

If you find yourself saying "and..." more than once, the paragraph is probably carrying more than one idea.

Example:

"Clear messaging helps customers understand what you offer, reduces confusion, builds trust, and improves visibility across platforms."

In this one sentence, look at the number of different ideas:

"This paragraph is about **clear messaging... and trust... and visibility.**"

And for a person who reads, this is normal. Humans don't have a problem understanding the message. However, **AI platforms that are making recommendations about business products and services,** have a harder time knowing what the message is really about, because there is more than one idea in the message.

Now lets improve the sentence for AI platforms and separate those ideas:

- Clear messaging helps customers understand what you offer.

- When customers understand you, trust increases.

- Clear messages also improve visibility across platforms.

Each sentence now handles one idea.

When each idea has its own space, confusion drops immediately—for readers and for AI systems trying to understand what your content is about.

What Is This Mainly About?

The one-sentence test works because it mirrors how AI actually reads.

When AI looks at a paragraph, it's trying to answer one basic question:

"What is this mainly about?"

To figure that out, it pays attention to a few things at the same time. To figure that out, AI checks whether the paragraph sticks to one topic and whether the sentences support each other. It also looks at whether the same idea appears more than once, and whether the paragraph could answer one clear question.

AI isn't checking these one by one.

It's looking at the overall pattern.

When all the signals point in the same direction, AI gets confident about what your content is about.

When they don't, it hesitates.

And when AI hesitates, your content is less likely to show up.

That's why separating ideas matters.

It removes uncertainty and makes your message easier to recognize.

Did You Know: AI Doesn't Read Between the Lines
Humans infer meaning from tone, context, and even what's left unsaid. AI doesn't. If you imply something without stating it, AI treats it as if it was never said. This reinforces why explicit, separated ideas matter — and it's something most people genuinely don't realize.

Clear Ideas vs. Mixed Ideas Table

This table shows focused ideas versus ideas that are unclear or mixed:

Clear Ideas	Mixed Ideas
One main point	Several points at once
Sentences support the same idea.	Sentences pull in different directions
Easy to summarize in one sentence.	Hard to explain simply.
Clear what the paragraph is about	Unclear what matters most
Easy for AI to recognize	Hard for AI to categorize.

Chapter 3 Key Takeaways

✔ One idea at a time makes your message easier to understand—for people and AI

✔ AI needs clearer structure than people do to recognize what your content is about

✔ Content Edges create clear boundaries around one topic

✔ When ideas stay focused, AI can categorize and recommend your content more confidently

✔ The One-Sentence Test helps you check if a paragraph carries too many ideas

Chapter 3 helped you break your writing into one idea at a time. Now Chapter 4 shows you the next step: say that idea in plain words.

Chapter Four

Say It Clearly

If you run a small business, you've probably seen the emails:

"Search has changed."

"SEO rules are different now."

"AI is rewriting how people find information."

These messages show up everywhere—marketing newsletters, industry blogs, and business articles. The advice is often detailed, sometimes urgent, and usually well-intentioned.

But after reading it, you often feel less clear—not more. And in AI search, vague language doesn't just sound polished—it gets skipped. If AI can't label what you mean, it can't bring you up as a recommendation.

What's missing isn't another tactic.

It's clear language.

Clear language is what makes your business easier to understand—and easier to identify in AI search.

Chapter 4 helped you break your writing into one idea at a time. Now we're doing the next step: say that idea in plain words. Because AI can't guess what you mean—and customers don't want to decode your message either.

Simple language is not unprofessional.

It's clear—and clarity helps your message travel.

Why Simple Language Works Better

People read quickly.

AI looks for patterns.

Both do better when your writing is straightforward.

Plain language:

- reduces confusion

- makes your point easier to grasp

- lowers overwhelm

- makes content easier to classify

This isn't about "dumbing things down." It's about making your work easy to understand in one reading.

And when your message is easy to label, it's more likely to show up when someone uses an AI tool to search for products, services, or recommendations.

Common Writing Patterns That Reduce Clarity

This happens for a practical reason: workplace language and AI discovery language are not the same.

At work, broad phrases often work because everyone already knows the context. Online, readers and AI tools don't have that context—so the same phrases can feel unclear.

This usually shows up as:

- big, vague phrases ("end-to-end," "solutions," "innovative")

- insider terms customers don't use

- one sentence trying to say four things

Example: "Leveraging strategic frameworks allows small businesses to optimize brand messaging in an evolving digital ecosystem."

A reader is left wondering: -**What does this actually mean? -What should I do with this?**

And AI runs into the same problem. It sounds professional—but it's not clear enough for AI search to recognize what you offer.

The Simple, Human Version

Here's the same idea, written plainly:

"Clear messages help small businesses explain what they do and connect with the right people."

Nothing important was removed.

But now it's easier to understand, remember, reuse, and recognize.

Quick Example: Same Product, Two Ways to Say It

Here's what this looks like in the real world—the same offer, said two different ways.

This kind of wording shows up all the time in product listings, "About" pages, and social posts:

Workplace-style (vague):

"I create high-quality, thoughtfully designed products for modern lifestyles."

That might be true—and it might sound professional—but it doesn't give enough information for AI search to know what you actually sell.

If a customer reads this, they still might ask:

"So... what is it?"

And AI runs into the same issue. It can't clearly label:

- what the product is

- who it's for

- what problem it solves (or what result it gives)

Now compare it to a version that gives those details:

AI-discovery (clear):

"I sell printable goal-setting journals for busy women who want a simple daily plan they can stick to."

Same creator. Same intention.

But now the message is specific enough to match real searches like:

- "goal setting journal printable"

- "daily planner for busy women"

- "simple routine planner"

When your wording answers what it is + who it's for + what it helps with, your message becomes easier for customers to understand—and easier for AI tools to recognize and recommend.

Now let's look at a few more examples side by side.

Workplace Language vs. AI Discovery Language

Now that you've seen the difference in one example, here are a few more.

Take a look at the Workplace vs. AI Discover Table below. The left column is the kind of language that works inside an office, where everyone already understands the context.

The right column is written for AI discovery, where your words have to stand on their own.

Notice how the AI Discovery version makes the message specific—so it's easier to understand, easier to categorize, and easier to recommend.

Workplace vs AI Discovery Table

Now let's use the same device in your sentence.

Practice: Rewrite for AI Clarity

Here's a simple rewrite exercise you can use on your own wording.

Step 1. Start with this sentence:

"Clarifying your content strategy allows you to optimize messaging and increase engagement across platforms."

Step 2. Rewrite it so a customer (and AI) can understand it in one reading.

Use this pattern:

I help (who) with (what) so they can (result).

Example:

"I help small business owners write a clear message so customers understand what they offer and take the next step."

Step 3. Reuse your Core Sentence (don't start over)

You've already written a version of this earlier in the book. Bring that sentence here.

If you don't have one yet, use this pattern now:

I help _____ **with** _____ **so they can**

_____ .

Now run it through this quick clarity check. Make sure it is:

- **short**

- **plain**

- **one idea**

You're not creating something new here—you're making sure the sentence you'll reuse everywhere is clear enough to carry forward.

If your sentence feels long or tries to cover multiple services, keep the same structure—but write **one version per offer** instead of blending them.

Chapter 4 Key Takeaways

✔ Simple language helps your message travel farther.

✔ AI can't guess what you mean—it needs plain, specific language.

✔ You don't need to sound impressive—you need to be understood.

✔ A clear sentence usually answers: who it's for, what you do, and the result.

You've learned to say your message clearly. Now Chapter 5 shows you how to make that message stick—by repeating it on purpose.

Chapter Five

Repeat On Purpose

How Repetition Builds Recognition—and Recognition Turns Into Customers

Lately, business headlines are full of phrases like AI agents and orchestrated workflows.

That's real. But most small businesses are working on a simpler problem:

Will the right customer recognize you after a few touchpoints—and will AI systems recognize you clearly enough to recommend you?

Most customers don't choose a business the first time they see it.

They choose when they recognize it.

Here's what that often looks like for a Tae Kwon Do school for kids.

Touchpoint 1: Notice

A parent scrolls Instagram and sees a post with the caption:

"Helping kids build confidence, focus, and respect—one class at a time."

They pause. It sounds relevant. Then they keep scrolling.

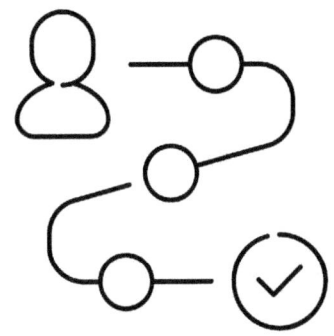

Touchpoint 2: Remember

A few days later, they see the school again—this time a short video of a shy student earning their next belt.

The message is the same:

"Helping kids build confidence, focus, and respect—one class at a time."

Same promise. Same tone.

Now the thought forms: *Wait... I've seen them before.*

Touchpoint 3: Choose

When the parent is ready to act, they don't just click a link.

They ask an AI tool:

"Best Tae Kwon Do school for kids near me"

or

"Martial arts program that helps kids with focus and confidence."

In that moment, the school's message has to do two things at once:

- Help the parent feel confident about choosing

- Help AI systems clearly understand what the business offers, who it's for, and what result it creates

When the message is consistent across posts, website, listings, and reviews, the business becomes easier to recognize—and easier to recommend.

This chapter shows how repetition builds recognition, and why recognition leads to enrollment, inquiries, and sales.

The Rhythm of Recognition

From a customer's point of view, **recognition** feels simple.

They notice you... forget you... then notice you again.

That second or third time is when something clicks:

"Oh yeah—I've seen them before."

Think about how a song grows on you.

It's rarely because you heard it once and loved it instantly.

It's because the same pattern repeats:

- the hook comes back

- the chorus repeats

- the rhythm stays familiar

The verse may change slightly, but you always know what song you're in. After a few listens, your brain works less.

The same principle works for business: repetition doesn't mean saying the exact same thing word-for-word—it means keeping your core message recognizable while the packaging changes slightly.

You can predict it. Name it. Remember it.

That's what repetition does in marketing too—not loud repetition, not constant novelty—just one clear message appearing in a few places until it becomes familiar.

The same principle applies to AI systems: repeated, consistent wording makes your business easier to label, categorize, and recommend.

What the Research Says

Research across psychology, usability, and marketing points to the same conclusion.

1) People remember information better when it's repeated over time.

Studies on the spacing effect show that information repeated across multiple moments is remembered more reliably than information delivered all at once.

Translation: One great post isn't a system. A clear message repeated across touchpoints is.

2) Familiarity increases comfort and preference.

The mere exposure effect shows that people tend to like and trust things that feel familiar—especially when they remain recognizable.

Translation: Familiar doesn't mean famous. It is easy to recognize.

3) People move faster when they can recognize rather than decode.

Usability research shows people perform better when information is easy to spot and understand without mental effort.

Translation: If someone has to work to figure out what you mean, many won't.

4) Repetition improves memory in marketing when it's spaced.

Advertising research confirms that repeated messages, spread out over time, improve recall compared to tightly packed repetition.

Translation: You don't need 50 messages. You need one clear message repeated calmly.

Research across psychology, usability, and marketing confirms what you probably already sense:

Why Repetition Matters Now (For Customers and AI)

Customers don't buy from the business that says the most.

They buy from the business they understand.

Most customers don't understand a business the first time they see it.

They see a post.

They skim a website.

They scroll past a listing.

They forget.

That's normal.

Repetition gives customers a second and third chance to understand without starting from scratch.

Now there's a second layer: AI systems are increasingly part of how customers discover businesses.

AI systems try to match questions like:

- "Who can help me with ___?"

- "What's the best ___ near me?"

- "What should I do if ___?"

They do this by looking for repeated signals:

- the same service described in similar words

- the same audience mentioned consistently

- the same outcome explained clearly

- the same categories appearing across pages and platforms

When those signals repeat, AI systems gain confidence.

And confidence increases the likelihood of recommendation.

If your message is scattered, customers may never see you—even if you're doing good work.

The Common Mistake: Always Trying to Sound New

It's easy to feel pressure to change your wording every time:

new hook, new angle, new identity, new voice.

But when wording changes too much, a business can accidentally sound like several different businesses at once.

Customers get confused.

AI systems become uncertain.

Repetition isn't boring when it's useful.

Repetition is how your message becomes recognizable.

Practice: Repeat Without Being Repetitive

This practice takes 15-20 minutes. If you're short on time, complete Steps 1-2 now and return for Steps 3-4 later.

The goal is simple: **one clear message, used intentionally in different places.**

You already created your core message earlier in the book. Bring it here. You're not starting over—you're making that message reusable across the places customers *and* AI systems actually check.

Step 1 — Retrieve your core message (one sentence)

Copy the sentence you already wrote:

I help (who) **with** (what) **so they can** (result).

Before adapting it, do a quick clarity check:

- **plain**

- **specific**

- **one idea**

If your business has multiple offers, choose **one offer** for this exercise. You can repeat the process later for others.

*Note: Your "core message" isn't limited to describing your business overall. You can also use this same format to describe a **specific offer or product**—especially your best-selling one.*

Example: "I help (who) with (product/offer) so they can (result)."

The goal stays the same: one clear, repeatable explanation that customers and systems can understand quickly.

Step 2 — Shape the same message for three roles

You'll use the *same meaning* in three slightly different ways—based on what each surface/site is responsible for.

Website anchor version (primary) Clear, explicit, and human-readable.This is the version your website should be built around—especially on your homepage, service pages, and About section. It helps both people *and* AI systems understand what you do.

Social version (connection) Shorter and more conversational.This version keeps the meaning intact but fits places where tone and attention matter more than full detail.

Explicit confirmation version (AI-facing) Most literal and category-forward.This version names the offer and outcome clearly so systems can summarize, categorize, and compare your business accurately. It's useful in listings, profiles, directory descriptions, and structured sections of your site.

The message does not change.Only the **shape** changes.

Step 3 — Place it where customers decide

Use your message where a real person is most likely to pause and decide, "Is this for me?"

- **Homepage** (opening section)

- **Service / product page** (top section)

- **Social bio or pinned post**

Keep it simple: one clear message, easy to understand in one reading.

You don't need new messages everywhere. You need **one steady message** where customers make decisions.

Step 4 — Place it where AI systems confirm

Now make sure the same meaning shows up in the places AI tools use to **verify** and **summarize** what your business is.

This is not about surveillance. It's about consistency across public business information.

Do a quick check of your **verification surfaces**:

- **Google Business Profile** (if applicable)

- **Primary directory / listing / marketplace profile**

- **Review platforms**

- *(Optional)* **About page** — if it's a primary source you share, include your **explicit confirmation version** in the first paragraph so your business is easy to recognize at a glance.

Your goal is alignment: the same category, the same offer, the same outcome—so systems can recognize your business without guessing.

How does this look in practice?

Example 1 — Service-Based Business

Core message (business-level)

I help small business owners clarify their messaging so customers understand what they offer and take the next step.

Where it shows up

- **Homepage (opening section)** Website anchor version appears as the main headline or subhead.

- **Service page** Same message, slightly expanded in the first paragraph.

- **Social bio** Shortened social version that keeps the same meaning.

- **Primary listing / profile** Explicit confirmation version used in the description field.

- **About page (optional)** The explicit confirmation version appears in the **first paragraph**, so the business is easy to recognize immediately.

Result: Customers understand what the business does quickly. AI systems can summarize and categorize the business without guessing.

Example 2 — Physical Product Brand

Core message (product-level)

I help people stay comfortable on their feet by designing breathable socks for long workdays.

Where it shows up

- **Homepage (opening section)** The website anchor version introduces the product and benefits clearly.

- **Product category page** The same message appears at the top, tied directly to the product line.

- **Social bio or pinned post** Shortened, conversational version.

- **Marketplace or directory profile** Explicit confirmation version names the product category and use case directly.

- **About page (optional)** An explicit confirmation version appears near the top to clearly state what the brand makes and who it's for.

Result: The product is easy to understand, easy to compare, and easy to recommend.

Example 3 — Business With Multiple Offers

Core message (one offer at a time)

I help first-time home buyers understand the mortgage process so they can make confident decisions.

How it's used

- This message is placed on the page for that specific service.

- Other services each get their **own** clear message using the same structure.

- The homepage introduces the primary offer first, not all offers at once.

Result: No blended messages. Each offer is clear, searchable, and recognizable on its own.

Marcus "The Business Coach" — Recognition Through Repetition

> Marcus is a business coach who helps first-time managers lead their teams. For two years, his bio changed every few months:
> - LinkedIn: "Leadership Development Specialist"
> - Website: "Executive Coach for Growing Teams"
> - Google Business: "Management Training Consultant"
> After reading Chapter 5, he standardized his message everywhere:
> **"I help first-time managers build confidence so they can lead their teams without second-guessing every decision."**
> **Within 90 days:** - When he Googled his business name + "what does he do," AI Overviews accurately described his work (before, it said "business consultant—various services") -
> Referrals started using nearly identical language:
> **-You help new managers, right?"**
> **- His LinkedIn connection requests mentioned his specific focus 3x more often**
> His visibility didn't increase because he posted more. It increased because people—and systems—could finally recognize him.

What These Examples Have in Common

- One clear message per business or offer

- Same meaning everywhere

- Different formats based on **role**, not creativity

- Explicit language placed early, where it's easy to find

You're not writing more. You're making what already exists **easier to carry forward**.

Chapter 5 Key Takeaways

✔ People choose what they recognize.

✔ Repetition turns "Who is this?" into "I know them."

✔ AI recommends what it can label with confidence.

✔ One clear message beats many scattered ones.

You've learned to repeat your message where it matters. Now Chapter 6 shows you how to carry that same message across different formats—without changing what it means.

Chapter Six

Show Your Expertise Clearly

Translation, Not Reinvention

By this point, you are not creating a new message. You are carrying the same message forward.

This chapter is about translation—taking one clear idea and expressing it across different formats without changing what it means.

Your expertise already exists. The goal now is to make it easy to recognize whether someone finds you through:

- your website

- an email

- a social post

- or an AI-generated summary

Different formats need different shapes.

They do not need different messages.

In this chapter, you'll learn how the same expertise can be:

- explained more fully in email

- organized clearly on your website

- shortened into quick, recognizable wording on social platforms

- and recognized by AI because your meaning stays consistent across the places it can access Nothing new needs to be invented.

Clarity is preserved by saying the same thing—calmly—in the form each channel requires.

What Expertise Really Means in Content

Expertise isn't something you add to your content.

It becomes visible when your message is easy to understand.

Expertise is not:

- sounding smarter than your customer

- using complex vocabulary

- listing credentials

- making bold claims

Expertise is:

- explaining what you do in plain language

- showing what you've learned from real work

- giving guidance people can use

- making the next steps more clear

When people understand you, they trust you faster.

When AI can clearly label what you do, it can recommend you more accurately.

Your job isn't to impress.

Your job is to clarify.

The One-Sentence Expertise Filter

Every expert can say what they do in one steady sentence:

"I help (who) with (what) so they can (result)."

This becomes your anchor across formats. It answers:

- Who is this for?

- What do they do?

- What result does it create?

Examples:

- **Clothing boutique:** "I help women choose everyday outfits so they feel confident and comfortable."

- **Fitness coach:** "I help busy women build strength so they feel better and stay consistent."

- **Marketing consultant:** "I help small businesses clarify their message so customers understand what they offer."

- **Beauty professional:** "I help clients simplify their routine so they feel put-together without stress."

When your expertise sentence stays consistent, your content becomes easier to recognize—for people and for AI.

This structure is also easier for AI systems to identify and categorize because it clearly states:

- the audience

- the action

- the outcome

Clarity is what makes expertise recognizable.

Quick worksheet prompt (60 seconds)

Fill in your anchor sentence:

I help _____ with _____ so they

can _____.

Choose Your "Source of Truth"

Clear translation starts with one decision: choose where your message will be explained fully first.

This is your source of truth—the place you can point to and say:

"This is what we do. This is who it's for. This is the result."

For most small businesses, that source of truth is your website (or main offer page) because it's public, structured, and built for decision making. It's where customers go to confirm:

- What do you offer?

- Is this for me?

- What happens next?

Once your message is clear there, your other content becomes easier to create—because it can support the same meaning instead of rebuilding it from scratch.

Quick example:

If you run a Tae Kwon Do school, your source of truth is the page that clearly explains: who it's for (ages), what happens (classes/training), what kids gain (confidence/focus), and how to start (trial class / sign-up).

If you don't have a website yet, your source of truth can be a single sales page, booking page, or pinned post—as long as it clearly explains your offer.

One Idea → Three Formats (Same Meaning)

In this chapter, "translation" means something simple: you take one clear message and express it in the format each channel needs—without changing what it means.

Different channels have different jobs along the customer journey:

- **Website:** where people confirm what you offer and what to do next

- **Social:** where people notice you quickly and begin remembering your message

- **Email:** where you can add context and build confidence after someone shows interest

Step 1: Start with one sentence that stays steady everywhere

This is your anchor—your clearest description of what you do.

Core expertise sentence (anchor):

"I help small business owners clarify their message so customers understand what they offer."

Step 2: Express the same meaning in three formats

Website (organized + decision-ready):

Websites are where customers slow down, scan, and decide. This is your clearest public explanation.

Example:

"Clear messaging for small businesses. I help you simplify what you do so customers understand your offer and feel confident choosing you."

Social (short + recognizable):

Social posts are quick, public reminders. Their job is to help people recognize your message fast.

Example:

"If people can't explain what you do, they can't choose you."

Email (more context + confidence-building):

Email lets you explain the same message with a little more detail—especially for people who are already interested.

(And while AI tools aren't reading private emails, email often sends customers back to your public pages when they're ready to decide.)

Example:

"Most marketing problems aren't strategy problems—they're clarity problems. When people can't quickly explain what you do, they hesitate. I help small business owners simplify their message so the right customers recognize them faster and feel confident taking the next step."

Same idea. Three shapes. One steady message.

A Real Example of Expertise Made Visible

A bakery owner I worked with didn't think of herself as an expert.

She told me, "I never went to culinary school, so I don't feel qualified to say much."

But when she described her work in one clear sentence—

"I help families celebrate moments by baking fresh, made-to-order desserts"—

everything changed.

People understood her faster.

They knew what she made, who it was for, and why it mattered.

Her posts performed better—not because she became more skilled overnight, but because her message became easier to recognize.

Nothing about her ability changed.

Only the clarity of her words did.

And that's what made her expertise visible.

The Compression Rule: Same Meaning, Fewer Words

When you move from a longer format (like a website page or email) into a shorter one (like a bio or social post), you can't bring everything with you.

That's normal.

The goal is to keep the meaning—even as the words get shorter.

Keep:

- who it's for (the audience)

- what you help with (the problem / need)

- what they get (the outcome)

Leave out:

- the full explanation

- the stories and examples

- extra details that make it longer, not clearer

Shorter doesn't mean weaker.

It means you're carrying the main point without all the background.

Quick check:

If the shorter version no longer tells people what you do and who it's for, you compressed too far.

When "Translation" Slips into "Rewriting"

If your message changes by platform, reset. Go back to your core sentence and shape everything from there.

The Quick Check Before You Publish

Before you post, send, or update a page, pause for ten seconds and ask:

- Is this still the same message as my core expertise sentence?

- Did I make it clearer—or did I just make it different?

- If someone saw this on a different platform, would they recognize it as me?

If you can say yes, you're good.

That's your message staying steady—even while the format changes.

Chapter 6 Key Takeaways

✔ Expertise does not change across platforms—format does.

✔ Translation preserves meaning; reinvention creates confusion.

✔ Your website (or main offer page) is your source of truth.

✔ Email adds context for interested readers; social builds **recognition**.

✔ AI learns from consistent public wording.

✔ One clear message is enough.

By now, your message is consistent across formats. The next step is understanding how AI confirms that consistency—and why some businesses get recommended more often than others.

Chapter Seven

Recognition in Practice

Before we explain how AI evaluates trust, let's see what clarity alignment —organizing your message so it's consistent across platforms—looks like in practice. VadoFilms is a creative media brand—very different from most small businesses—but the principles apply universally. Here's what changed, and why it worked.

Case Study

How Reorganizing an About Page Made a Creative Brand AI-Readable

Venus 'Vado' Teguia Kouhm

VadoFilms LLC is a long-standing   creative media brand founded in 2011, led by an award-winning director with a broad portfolio spanning music videos, film, and visual storytelling. Over time, the brand accumulated a wide range of assets, accomplishments, and public references across multiple phases of work.

Rather than treating visibility as the problem, the review began with recognition—specifically how clearly the business could be identified as the same entity across its key reference points. The About page became the primary focus because it acts as a central confirmation source within the broader verification infrastructure.

While AI systems already surfaced VadoFilms in connection with awards and directing, the About page presented roles, achievements, and history in dense blocks of text. This made important trust signals harder to distinguish and slowed both human understanding and system-level confirmation.

No new content was added.

No platforms were changed.

No optimization tactics were introduced.

Instead, the work focused on clarity alignment—organizing existing information so it could be more easily recognized and confirmed.

Awards were separated and surfaced as distinct trust signals rather than embedded in a long narrative. Major accomplishments were linked to dedicated pages, reducing overload on the About page while improving consistency across reference points. Language was adjusted to reflect the brand's current focus, and legacy service descriptions that no longer represented active offerings were de-emphasized.

The result was not a change in credibility, but a change in consistency. The business already had the signals—they were simply competing with each other.

The Founder later described the shift using a simple metaphor: "everything was already there—it just wasn't where it was supposed to be."

Once information was organized and aligned, the About Page on the website became easier to read for humans, and easier for AI to scan and confirm information.

This review was like adjusting a camera lens: Nothing new entered the frame—but once focus was set, the business became easier to see and easier to describe consistently. The client called it a "level up," like any technology transition: businesses that don't update how they show up online can get left behind. By improving structure, formatting, and consistency across the verification infrastructure, the business became easier for AI systems to categorize, summarize, and recommend.

Nothing new had been created during or after the clarity alignment.

But consistency had been restored.

With recognition stabilized, next-step decisions became easier—because the business was consistent across key reference points.

This is what clarity alignment looks like in practice: not adding new content, but organizing what already exists so systems can confirm it. The question is: why does this work? Chapter 9 explains the trust signals AI uses to evaluate businesses—and why being clearly identifiable has to come before optimization.

Chapter 7 Key Takeaways:

✔ Recognition often doesn't require new content—just better organization

✔ Trust signals work best when they're easy to distinguish, not buried in dense text

✔ Clarity alignment focuses on consistency, not expansion

The chapter 7 case study showed what clarity alignment looks like. **But why does organizing information make such a difference?** Chapter 8 explains how AI evaluates trust—and why consistency matters more than claims.

Chapter Eight

What AI Trusts

Why Consistency Matters—and How Optimization Works

AI systems don't simply surface content. They evaluate legitimacy. Before recommending, summarizing, or prioritizing a business, AI systems review content for confirmation that the business is real, stable, and consistently identifiable across sources.

If optimization in AI-driven search and recommendation systems has felt unpredictable, it's often because AI could see your content—but couldn't confidently recognize your business.

How AI Understands Trust (In Human Terms)

AI does not "believe" claims.

It cross-checks patterns.

When an AI system encounters your business, it compares what you say about yourself with what appears elsewhere—your website, bios, directories, listings, reviews, and brand references. When those sources align, your business becomes easier to identify, categorize, and summarize accurately.

When they don't align, the system hesitates.

If your website calls you a coach, your profile calls you a consultant, and a directory lists you as an HR professional, the system has no stable version to confirm. It isn't judging which description is better—it simply can't determine which one is consistent enough to carry forward.

In this chapter, **trust means recognizability**. Not authority. Not popularity. Not persuasion.

Trust is the system's confidence that it understands *what kind of business you are*—and can describe you the same way wherever you appear.

What Are Stable Categories?

Stable categories mean your business is classified the same way everywhere it appears online—and that classification doesn't keep changing.

A category answers the question: "What kind of business is this?"

AI uses categories to decide when, where, and whether to show or recommend you.

Examples of Stable vs. Unstable Categories

Which column below looks more like your business?

This table shows how the same business can appear clear or confusing depending on how it's listed on different platforms—and whether its category stays consistent across key Recognition Surfaces.

What AI sees:

- **Stable:** one clear pattern → easy to recognize and match

- **Unstable:** mixed categories → AI has to guess and often slows down or becomes inconsistent

If you want optimization to stick later, stabilize your category first.

When categories don't match, AI has to guess which one is accurate, so your visibility can become inconsistent.

Why Stable Categories Matter to AI

Stable categories help AI:

- group your business correctly

- match your business to relevant potential customer searches and recommendations

- summarize what you do accurately

- decide where you belong (maps, local results, knowledge panels, AI summaries)

When categories keep shifting, systems can't reliably confirm what kind of business you are.

Authority Without Performance

Many businesses try to signal authority through volume: more posts, more claims, more positioning.

AI does not reward performance.

It rewards clarity.

Specific language, clear roles, and restrained positioning are easier to classify and trust than constant reinvention. Authority is demonstrated when a business sounds the same over time—not louder. One clear anchor compounds trust faster than scattered effort.

This means saying, "I help small businesses clarify messaging' every time—not 'fitness guru' one week and 'brand strategist' the next."

Where Optimization Fits—and Why It Often Fails

Optimization means making it easier for people and AI to find you at the right time.

But optimization is not step one. It's step three.

First, your business has to be easy to recognize. That means you describe what you do the same way in the places people and AI check—like your website, any online business Profile, and your main bios.

Second, when those details match, AI can confirm your business is consistent across sources. That confirmation is what we call "trust" in this chapter.

Only then does optimization work the way you expect.

Here's the simple order:

- **Recognition** — AI can tell who you are

- **Trust** — AI can confirm your business is consistent across sources

- **Optimization** — AI shows you more often, in the right situations

When people optimize too early, everything feels inconsistent. You change keywords. You rewrite your bio. You update listings. You post more. But nothing sticks.

Not because you're doing it wrong—but because the patterns don't match yet.

If one page says you do one thing, and another page says you do something else, AI has to guess. And when AI has to guess, it usually won't recommend you confidently.

That's why Clarity Alignment comes first to make your effort actually work.

One Of The Ways Business Information Travels

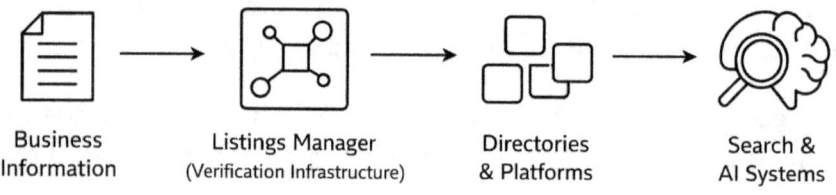

Business Information → Listings Manager (Verification Infrastructure) → Directories & Platforms → Search & AI Systems

Verification Infrastructure, Not Hacks

There are systems whose primary job is to keep business information consistent across the internet. These systems are called listings managers. You don't usually sign up for them directly. You may not even know one is involved with your business. But search platforms, AI systems, and directories interact with them all the time.

Directories, listings managers, and review platforms play an important role—but not the role they're often assigned. These systems function as verification infrastructure.

Think of verification infrastructure like the systems behind a credit report or a background check. You don't interact with them directly.

Credit scores work because information about you comes from many independent sources. You don't set your own score, and you don't report your own history. Instead, lenders report the same details over time, and credit agencies compare those reports to confirm a consistent picture. When the information matches, confidence increases. When it conflicts, the system becomes uncertain or flags the record.

Verification infrastructure works the same way for businesses. It's the network of directories, listings managers, review platforms, map services, and brand references that repeat and confirm the same business information across the internet. They help distribute and cross-check consistency across reference sources, but they do not create clarity on their own.

For a deeper look at how listings managers work and how to update your information through them, see Appendix A: Understanding Listings Managers

That's what makes optimization work.

Chapter 8 Key Takeaways:

✔ Trust means recognizability—not authority or popularity

✔ Stable categories help AI group, match, and summarize your business accurately

✔ Recognition comes first, then trust, then optimization

✔ Verification infrastructure supports recognition but doesn't replace clarity

You understand how AI builds trust. Now Chapter 9 shows you how to optimize—once recognition is stable.

Chapter Nine

Optimization That Sticks

The steps in this chapter are not a personal preference—they reflect how modern AI and decision systems actually interpret information.

GEO: Why This Work Has a Name Now

In Chapter 1, we called this AI Visibility and mentioned it's sometimes called GEO. Here's why that term matters in industry research—and what it means for how you approach visibility.

In industry research, it's increasingly described as Generative Engine Optimization (GEO)—same concept, technical name.

GEO reflects a change in how visibility works when AI systems generate answers instead of displaying lists of links. Rather than ranking pages, AI systems synthesize information across sources and surface businesses they can clearly understand, confirm, and describe.

According to research and analysis from McKinsey & Company, AI-driven search now acts as a decision layer. It compares information across websites, listings, reviews, and reference points, then generates responses based on what it can interpret confidently. In this environment, content is rewarded when it is clear, structured, and consistent enough to be cited.

This is what GEO focuses on:

- making meaning easy to extract

- reducing ambiguity across sources

- aligning descriptions so AI systems can confirm them

- supporting synthesis, not just discovery

Importantly, GEO is not about chasing new tactics. It is about being understandable in a system that summarizes instead of browses.

That's also what you've been building toward in this book. The Topic Trio, the Question Bank, Content Edges, consistent language—all of it prepares your business to function well in a GEO environment.

This order isn't arbitrary. Research from Stanford and MIT shows that AI systems interpret structured, consistent information more reliably. When you confirm before you expand, reduce before you add, and strengthen before you select, you're working with how these systems actually process information—not against it.

You now have the foundation. This chapter shows you how to put it to work.

How to Improve Visibility Once Recognition Is Stable

Up to this point, you have been stabilizing your foundation. You clarified what you do. You narrowed your message. You aligned how your business appears across the places people and AI systems use to verify legitimacy.

That work makes everything else easier. Now you move into action—calmly and selectively.

In this book, optimization means improving how easily your business can be confirmed and matched when someone searches. It works when recognition is already stable—meaning three things are true:

1. You can describe your business in one clear sentence.

2. That sentence matches across your main recognition surfaces.

3. Your business category is stable and does not change by platform.

When those conditions are in place, optimization becomes a downstream accelerator. People understand you faster. AI systems describe you more accurately. Your business shows up in the right moments with less effort. And a few well-chosen updates create more impact than constant changes.

If those conditions are not in place, optimization tends to feel unpredictable—not because you're doing it wrong, but because the system cannot confirm you yet.

The Order That Makes Optimization Work

Optimization happens in a simple sequence:

1. Confirm

2. Reduce

3. Strengthen

4. Select

Each step makes the next one easier. And each step increases the value of your effort without adding more noise.

Step One: Confirm Your Anchor Is Working

In Chapter 8, you chose your anchor surface (your main reference page). Now you're going to confirm it's doing its job.

Your anchor is "working" if a potential customer can answer these three questions after a quick scan:

What does this business do?

Who is this business for?

What result does this business help create?

If the answers are clear, you're ready to continue.

If they feel unclear, make a small edit on your main reference page until a potential customer can confirm the meaning in one quick scan.

The goal is simple: make your message easy to confirm in one pass.

Step Two: Reduce the Places That Create Confusion

With your main reference page confirmed, the next step is simplification.

You are looking for the small number of places where your business is described in a way that creates mixed signals. When those places are aligned, people understand you faster—and AI systems can confirm you with more confidence.

Scan for the four most common confusion points:

☐ Category: Is your business labeled the same way everywhere?

☐ Service or product name: Do you call the same offer by different names?

☐ Old services: Are past offerings still listed as current?

☐ Audience: Does your "target market" change across platforms?

Now take action:

☐ Select the three items above that most affect how your business is understood

☐ Update those three so the wording and category match your main reference page

☐ Stop once those are aligned

You don't need to address everything at once. Clearing the biggest points of confusion often creates the most immediate improvement.

Step Three: Strengthen Existing Trust Signals

This step improves the quality and clarity of what already exists. You are not scanning for new issues or choosing where to focus long-term. You are making a small number of refinements that help people and AI systems confirm your business more easily.

Choose two or three actions from the list below:

☐ Improve headings or section labels so your category and service are easier to identify

☐ Add a short summary paragraph to your main reference page (top of page or first scroll)

☐ Update the language in one high-visibility listing so it matches your core wording

☐ Adjust the wording of one review response so it reflects the same message used elsewhere

☐ Remove or de-emphasize outdated services or descriptions that no longer represent what you offer

Complete this step:

☐ Confirm the wording now clearly communicates what you do, who it's for, and the result

☐ Stop after two or three improvements are complete

This step strengthens trust by making your business easier to read, summarize, and confirm—without expanding your footprint.

Step Four: Select Your Confirmation Focus

By now you've confirmed your anchor and reduced confusion. This step decides where you'll maintain that clarity going forward.

You are not editing broadly here. You are deciding which two places matter most for confirmation.

Start by answering this question: Where do people go to confirm your business before choosing it?

Now choose two locations:

☐ Primary confirmation location (the main place you want people to check) Primary: _____

☐ Supporting confirmation location (the second place that reinforces the same message) Supporting: _____

Complete the focus decision:

☐ Confirm both locations clearly communicate what you do, who it's for, and the result

☐ Treat these two locations as your ongoing maintenance pair

☐ Use them as your reference point before making future updates elsewhere

When these two places stay aligned, confirmation becomes faster—and everything else becomes easier to manage.

The Simple AI Check in Action

Jennifer owns a boutique fitness studio specializing in strength training for women over 40. She'd been in business for 5 years with a loyal client base—but new customers rarely found her online.

She ran the Simple AI Check from Chapter 9:

- **Opened ChatGPT and asked the question: "What does Jenny's Fit & Formation do?"**

- **ChatGPT Result:** "It appears to be a fitness studio, but I don't have enough consistent information to provide details."

That was her wake-up call. She spent 2 hours aligning her message across her Website homepage, her Google Business Profile, Instagram bio and Yelp listing.

This is Jennifer's new core message for her boutique that she put everywhere:

"I help women over 40 build strength so they feel confident, capable, and injury-free."

30 days later, she ran the AI Check again:

- **ChatGPT 's response:** "Jenny's Fit & Formation is a fitness studio in California that specializes in strength training for women over 40, focusing on building confidence and preventing injury."

Jenny didn't create new content. She aligned what already existed.

Within 60 days. she had:

- 5 new clients mentioned finding her through "AI search".

- Her Google Business Profile views increased by 35%.

- Referrals started using her exact language: "You help women our age, right?"

How to Tell If It's Working

You don't need complex tools to check whether your optimization is taking hold. You just need a clear way to see whether your business is being summarized accurately.

Open an AI tool and ask:

"What does [Business Name] do?"

"Who is it for?"

"What does this business help people achieve?"

Then compare the response to your main business sentence.

If the summary matches your message, your wording is consistent enough for an AI system to describe you clearly. That's a strong sign your optimization is working.

If the summary feels off, treat it as useful feedback. Return to your public-facing pages and check for the most likely causes:

a category mismatch

outdated services still listed

different descriptions in different places

language that shifts by platform

Make one adjustment that improves consistency, then run the check again. This is how optimization becomes steady: confirm, adjust, reinforce.

Don't expect immediate changes. AI systems reflect clarity improvements gradually, as they revisit and compare information over time. If your message is consistent, the summary will catch up.

What to Expect After You Make Updates

AI systems do not update on a fixed schedule. Instead, they reflect changes gradually as information is revisited and reconciled across sources.

Clearer descriptions show up sooner. Reputation signals take longer. Check quarterly.

What "**is it working**" looks like:

AI summaries describe your business more consistently

Your category stops drifting across platforms

Your business becomes easier to explain—even by someone else

Optimization compounds quietly. Consistency over time produces more durable results than frequent changes.

What Optimization Is Not

Optimization is not:

-weekly bio rewrites

-constant keyword changes

-chasing platform updates

-adding channels before clarity compounds

-If optimization feels anxious or urgent, it's often a sign that the earlier steps need more time — not that you've done something wrong.

-Optimization that sticks feels quieter, not louder.

Chapter 9 Key Takeaways

✔ Optimization works only after recognition is stable.

✔ Order matters more than tactics.

✔ Reduce contradictions before adding effort.

✔ Strengthen what already exists.

✔ Show up where confirmation happens.

✔ If AI can summarize you accurately, optimization is already working.

Once your business is easy to recognize, optimization stops feeling like pressure and starts feeling like alignment.

The next question is no longer how to optimize. It's how businesses that understand this shift are preparing for what comes next.

Chapter Ten

The AI-Ready Business

By the time optimization is working, it becomes less noticeable.

Businesses that are well prepared for AI-driven discovery focus on stabilizing how they are understood. Their message holds steady across time and platforms, allowing recognition to build without constant adjustment.

AI systems recommend businesses they can recognize, confirm, and describe with confidence. When that confidence is present, visibility follows naturally.

This chapter focuses on what becomes true once clarity is already in place—and how that state shows up in real market behavior.

That difference is the difference between optimization and readiness.

From Optimization to Readiness

An AI-ready business has signals that are stable enough to be trusted across systems.

In practical terms, readiness shows up when:

- Your business can be described consistently without you being present

- Your offerings can be categorized without explanation

- Your expertise can be summarized without distortion

When these conditions exist, AI systems do not need to interpret or infer intent.

They can confirm what is already clear.

This pattern is already visible in how current markets and consumers behave.

Those conditions show up first in how certain businesses operate day to day.

What the Data Is Actually Pointing Toward

Industry research shows a widening gap between businesses that are active and businesses that are consistently discoverable.

Consumer behavior studies, including research from National Retail Federation, indicate that buyers increasingly rely on aggregated answers rather than individual sources. Instead of comparing multiple websites manually, people ask systems to compare on their behalf.

At the same time, broader operational signals point toward consolidation:

- Fewer channels are being prioritized

- Systems are favored over short-term campaigns

- Clarity is valued more than novelty

As the number of available options increases, the work of evaluation shifts. Rather than exploring every option directly, potential customers rely on systems to narrow, summarize, and validate choices for them. This is what researchers call **cognitive load reduction** — AI-powered search is designed to give users fewer choices and more synthesized answers so decision-making requires less effort. The tradeoff is that interpretation moves upstream, before humans engage.

In that environment, confirmation becomes more valuable than discovery. Systems favor options they can verify and align across multiple sources without contradiction.

What matters is not being one of many visible options, but being the option a system can confidently recognize, verify, and describe.

Those conditions show up first in how certain businesses operate day to day.

A Consistent Pattern Among Early Adopters

Among businesses adapting well, a consistent behavioral pattern emerges.

They change their messaging less often.

They make fewer reactive adjustments.

They spend less time correcting misunderstandings across platforms.

Their attention is centered on whether what already exists is easy to understand and confirm, rather than on producing more output.

As a result, optimization becomes selective and bounded. Once confirmation is established, effort decreases rather than escalates.

From the outside, this stability often reads as confidence.

From the inside, it registers as relief.

What Do AI Systems "Remember" About You Over Time?

These systems work by comparing the public information your business already puts out — your website, your listings, your reviews — and checking whether it tells a consistent story.

Think of it like this: every time your business shows up online—on your website, in a listing, in a review—AI systems take note. Not in a creepy way. In a pattern-recognition way.

Your public information stays online. It gets indexed. And when someone asks a question your business could answer, AI pulls from those sources to build a response.

Here's the key part: **AI doesn't start from scratch every time.**

When AI encounters your business again, it's not meeting you for the first time. It already has a working understanding based on what it's seen before. If your message has been consistent—same category, same wording, same offer—AI recognizes the pattern immediately and moves forward with confidence.

If your message keeps changing? AI has to pause, compare, and reconcile the differences. Which version is accurate? What does this business actually do? That hesitation slows everything down.

Over time, AI builds what you might call a **composite picture** of your business: how you're categorized, what you're known for, and how reliably that information aligns everywhere you appear.

New information doesn't erase the old picture—it either strengthens it or complicates it.

When your information stays consistent, AI can confirm what it already knows. No guessing. No reconciling contradictions. Just confidence.

When it shifts frequently, AI has to work harder—and uncertain systems recommend less often.

Recognition compounds through steady repetition, not sudden bursts of activity.

And here's the advantage for small businesses: you can stabilize that picture faster than a large company can. Fewer layers. Fewer decision-makers. Faster alignment.

Why Smaller Businesses Often Adapt Faster

Smaller businesses often adapt more smoothly to AI-driven discovery because alignment can be achieved with fewer layers of decision-making.

Positioning decisions typically pass through fewer stakeholders, allowing messages to settle and remain consistent once they are set. With fewer offerings and fewer digital properties to manage, the same descriptions tend to appear wherever the business shows up.

Humans and AI systems encounter a stable explanation across time and platforms, making recognition and confirmation easier to establish. As a result, clarity builds faster and is easier to maintain over time.

That advantage holds regardless of how specific tools or platforms change.

Future Positioning Without Prediction

You don't have to predict where AI is going.

You have to stay consistently understandable.

AI systems surface businesses they can interpret and describe with confidence across sources. People increasingly rely on summarized answers and let systems compare options for them.

In that environment, clarity holds its value even as tools and platforms change.

When those conditions are met, readiness becomes visible in everyday operations.

What Readiness Looks Like in Practice

Readiness shows up as stability.

When your business is easy to recognize and describe across platforms, visibility no longer requires constant protection. Fewer updates feel urgent. Explanations become simpler. Effort becomes more selective.

That is the quiet advantage. Once understanding holds steady, you can move forward without restarting your message each time the environment shifts.

Chapter 10 Key Takeaways

By this point, the conditions for readiness are clear.

- You understand how recognition stabilizes, how systems remember meaning over time, and why clarity reduces effort rather than increasing it.

- What remains is not more explanation, but application.

The next chapter translates this understanding into a simple, bounded plan. Not to create readiness—but to work from it. Chapter 11 focuses on how to move forward once clarity is already doing its job, using a short, steady horizon that prioritizes consistency over urgency.

Chapter Eleven

Your First AI-Readable Article

You've learned the principles. Now it's time to build something real. This chapter gives you a step-by-step process to create one AI-readable article you can reuse everywhere.

The purpose of this activity is to create a single, stable explanation of what your business offers—written clearly enough that both customers and AI systems can understand, summarize, and reuse it accurately over time.

By the end of this chapter, you will have:

- one real customer question to anchor your content

- one clear answer that defines your business or offering

- one structured article that reinforces the same meaning across sections

This article becomes a reference point. Other content can echo it, link to it, or shorten it without changing the message.

You are establishing clarity that can be carried forward.

Why Systems Matter Before You Write

AI systems follow systems. They build understanding by comparing information across sources and across time. When your business is explained the same way—using stable language and a clear structure—systems can recognize what you do, categorize it accurately, and summarize it without needing to reinterpret your meaning each time.

This is why consistency matters more than volume. One clear explanation, held steady, is easier for systems to carry forward than many variations of the same idea.

The method used in this chapter is designed to support that system behavior.

Big Picture: The H.A.R.P. Method

You've been applying these principles throughout the book—now we're naming them so you can reuse them intentionally. H.A.R.P. is the standard this book uses for creating content that works with AI systems rather than against them. As you build your first article, you're practicing four conditions that make your message easier to understand, summarize, and repeat.

This book uses the H.A.R.P. method as its organizing framework for AI-readable content.

H — Human-First

Write the way you would explain something to a real customer.

Use simple language. One idea at a time. Short, natural sentences.

A — AI-Readable

Make the meaning easy to interpret.

Use clear headings, focused sections, and consistent wording so your content can be categorized and summarized accurately. This chapter will walk you through how to do this step by step.

R — Repeatable

Repeat your core language on purpose.

Consistency helps both people and AI systems remember what your business is and what it offers.

P — Practical & Helpful

Answer a real customer question clearly.

Be specific. Be useful. Help the reader understand something without extra explanation or filler.

In practice, H.A.R.P. looks like this:

- Choose a real question

- Write one clear answer

- Structure the answer so it's easy to follow

- Keep your wording consistent

- Confirm it can be summarized accurately

Clarity makes your content easier for people to understand—and easier for AI systems to recognize and recommend.

How Content Edges and H.A.R.P. Work Together

(Why the article you just wrote actually works)

Here's what just happened—and why it matters.

You built something AI systems can actually use. Not because you followed a formula, but because you gave them what they need most: **clear boundaries and clear explanations.**

Think of it like building a room. Content Edges are the walls. H.A.R.P. is everything inside.

Content Edges: Drawing the Line

A Content Edge answers two simple questions: **"What is this about? — and — what is it *not* about?"**

When you chose one customer question and wrote one focused article, you drew that line. You didn't try to explain your entire business, answer five questions at once, or sneak in three different topics. You stayed inside the edge.

That's why AI could follow you.

H.A.R.P.: Making the Inside Clear

Once the boundary is set, H.A.R.P. shapes what happens inside it:

- **Human-first language** keeps it readable

- **Clear structure** makes it easy to scan

- **Repeatable wording** keeps the meaning stable

- **Practical takeaway** keeps the point intact

Together, Content Edges and H.A.R.P. send AI a clear signal:

"This content answers one specific question. It stays focused. You can summarize it without guessing."

And that's exactly what AI needs.

What AI Is Actually Checking

AI systems aren't judging creativity or effort. They're checking three things:

1. **Does the topic stay focused?**

2. **Is the explanation structured?**

3. **Can this be summarized accurately?**

When the answer is yes to all three, your content becomes easier to categorize, reuse, and recommend.

This is why one clear article outperforms ten scattered ones.

It's not about writing more. It's about writing in a way systems can carry forward.

In the next chapter, you'll learn how to hold this same explanation steady across time and platforms—so recognition has a chance to build.

Your First AI-Readable Article

A Step-by-Step Guided Build

This build takes 25-35 minutes. Want to see the finished product first? Skip to the restaurant example below, then come back to build your own. **If you're short on time, read through once,** *then complete it later*

Follow the steps in order. You are building one article, not completing a worksheet.

Step 1: Choose a Real Customer Question

Purpose: Pick a question customers ask when they are deciding whether your product or service is a fit.

Start with a real customer question. **Use the Question Bank you created in Chapter 3 and select one question from that list, or:**

If you're just getting started, search for questions in:

- customer reviews

- competitor reviews

- social media comments

- FAQ sections for similar businesses

Use the checklist below:

☐ Choose a question a customer would naturally ask

☐ Make sure the question reflects a customer decision, not a marketing goal

☐ Keep the topic narrow enough to answer clearly in one article

Example question (restaurant owner):

What kind of food does this restaurant serve?

Screener Check (why this works):

☐ Reflects how customers actually search or decide → Cuisine type is a primary filter customers use.

☐ Makes clear what the customer is trying to figure out → They want to quickly categorize the restaurant.

☐ Narrow enough to answer in one article → One article can define the style, signature dishes, and who it's for.

If an AI platform cannot answer this question confidently, it cannot recommend the business.

Step 2: Define the Core Answer

Purpose: Write one sentence that clearly explains what you offer, using the same words you want customers—and AI systems—to repeat about your business.

Example core answer (restaurant owner):

"This restaurant serves Southern-style comfort food with a focus on slow-cooked meats, classic sides, and family-style portions."

Instruction:

Using the question you chose in Step 1, write a one-sentence answer to open your article. This sentence becomes the anchor for everything you write next.

Use this checklist:

☐ One sentence

☐ Uses repeatable language

☐ Specific, not generic

Screener Check:

☐ Clear in one sentence

☐ Uses repeatable language

☐ Specific enough to categorize accurately

Stop here when:

You can read the sentence out loud and someone immediately understands what you offer and what makes it distinct.

Step 3: Organize the Article Around the Answer

Purpose: Organize the article so each section explains one part of your answer.

☐ Write 3–5 section headings

☐ Each section covers something different

☐ Sections explain rather than sell

Here's what a complete AI-readable article looks like using this structure. Notice how each section answers one part of the core question.

Restaurant

Article Example

Here's a short article for the restaurant owner in the example above. The article is written based on the clear, AI-Readable structure shown in the table above. Notice how the sections in Step 3 become the section headings for the article. The writing is intentionally simple, explanatory, and easy to summarize.

What "Southern-Style Comfort Food" Means

Southern-style comfort food is built around familiar, home-style dishes prepared with time and care. Meals are designed to feel hearty and satisfying, drawing on traditional recipes that emphasize flavor, slow preparation, and generous portions.

This style of cooking focuses less on trends and more on consistency—food that feels recognizable, filling, and meant to be shared.

Signature Dishes and Cooking Style

The menu centers on slow-cooked meats such as braised chicken, pulled pork, and beef cooked low and slow to develop rich flavor. These dishes are paired with classic sides like collard greens, cornbread, mac and cheese, and seasoned vegetables.

Cooking methods prioritize patience and depth of flavor rather than speed, resulting in meals that feel substantial and comforting.

Who the Menu Is Designed For

This restaurant is designed for people looking for familiar, satisfying meals rather than small plates or experimental cuisine. Portions are generous, and many dishes are suited for sharing, making the menu appealing to families, groups, and anyone seeking a traditional dining experience.

The food is approachable and filling, meant to leave diners satisfied rather than curious.

How This Restaurant Differs from Similar Options

While many restaurants offer Southern-inspired dishes, this restaurant focuses specifically on classic comfort food prepared with consistent methods and traditional sides. The emphasis is on slow cooking, familiar flavors, and meals that feel complete rather than reinvented.

That focus makes it easy to understand what the restaurant offers and who it is for.

Step 4: Use Consistent Language

Purpose: Use the same language throughout so your meaning stays clear everywhere it appears. Be consistent.

☐ Use the same category terms throughout

☐ Keep cuisine, style, and audience descriptions stable

☐ Match section headings to body text

Avoid rotating synonyms unless the distinction truly matters.

Step 5: Summarize the Article

Purpose: Write a 1–2 sentence summary to test clarity. You don't publish this—use it to confirm your article says one clear thing.

☐ Write a 1–2 sentence summary

☐ Confirm it matches your core answer

☐ Remove anything that introduces a new idea

Example summary:

"This restaurant serves Southern-style comfort food, specializing in slow-cooked meats and traditional sides. The menu focuses on familiar, hearty dishes designed for sharing."

If the summary matches the core answer, the article is Human-First and AI-Readable.

What to Take Forward

You don't need many articles like this.

You need a few that:

- answer real questions

- use consistent language

- maintain a stable structure

That's what allows clarity to compound—across platforms, over time. This article now becomes a stable explanation you can reuse, reference, and repeat.

One clear article is powerful—but only if you protect it. Chapter 12 shows how to keep your message steady for 90 days so recognition has time to build.

Did You Know: AI Reads Your Page Top to Bottom — and Stops Early

Most people don't know that AI systems weigh the first few sentences of a page more heavily than what comes later. If your opening paragraph mixes three ideas, AI may categorize you based on a confused first impression — even if the rest of the page is clear. This connects directly to the one-idea principle and gives the reader a concrete reason to care about structure.

Chapter Twelve

The 90-Day Continuity Plan for AI Visibility

A continuity plan is a short plan for keeping your core explanation consistent across the places your business already shows up.

During the 90 days, you focus on four things:

1. Keep your core explanation the same

If you completed the H.A.R.P. build in Chapter 12, you already created this. It's the one clear sentence you wrote (your "what we do" line), plus the matching wording you used in your bio or website and in your first article.

For the next 90 days, keep that core explanation unchanged.

Don't rewrite your category or switch to a new main message.

If you edit anything, only edit to match the same wording everywhere—not to change what you mean.

2. Repeat your key language on purpose

In Chapter 12, use the same phrases and words you selected to describe your business across pages, profiles, and posts. Familiar wording helps people and systems recognize you faster.

3. Focus only where clarity already exists

Select the places customers actually see first. Focus on your "front door" surfaces—the pages and profiles someone is most likely to check before they decide:

- your website About page

- your main service/product page

- your top online Business directories and profiles

- your social bio and your most active social page, any profile that shows up on the first page of search results for your business name

4. Notice what becomes easier for you

After 90 days, you may notice:

- people ask fewer "so what do you actually do?" questions

- you don't feel the urge to keep rewriting your descriptions

- your website, listings, and profiles sound like they belong to the same business

These are signs that customers and AI systems are recognizing your message consistently.

The Value of Continuity

By now, you've done the hard part.

You moved from scattered visibility tactics to a clear explanation of what you do. You learned how AI-driven search works, where your meaning gets lost, and how to make your business easy to recognize and describe. You built one strong piece of AI-readable content that can be repeated without changing the message.

Continuity is what makes all of that stick.

When your explanation stays consistent, your effort shifts:

- you spend less time correcting confusion

- you stop rewriting your message every time a platform changes

- you save time because your content is doing the explaining for you

- you can move forward without restarting

Chapter 10 showed what it looks like when a business is easy to confirm. Chapter 11 showed how to write in a way systems can understand. This chapter shows how to protect that clarity for 90 days so it can build.

If AI can't read you, it can't recommend you.

Chapter Thirteen

BONUS CHAPTER: Become Easy to Recognize Everywhere

90-Day Plan—Checklist

If you're ready to take immediate action—even before completing your 90-day plan—this checklist gives you a fast way to spot and fix the biggest visibility gaps.

A 10–15 minute checklist to reduce contradictions and stabilize understanding

Recognition is what makes your business easier to find, easier to confirm, and easier to choose. When your wording and category stay consistent across the places people and AI systems check—your website, listings, profiles, and reviews—your business becomes simple to identify and simple to summarize.

This checklist gives you a fast way to spot contradictions, align your key recognition surfaces, and walk away with a clearer, steadier public description you don't have to keep rewriting.

If you work with an agency or vendor, ask if they use a listings manager (see Appendix: Understanding Listings Managers for details).

Step 1: Identify Your Core Description

(This is your reference point—not your slogan.)

Write the clearest, simplest sentence that explains what your business does.

I help _____ with

_____ so they can

_____.

This sentence should be easy for a stranger—or a system—to repeat accurately.

Step 2: Check Your Recognition Surfaces

(These are the places people and AI systems check to confirm who you are.)

Review only the places below. You do **not** need to check everything online.

☐ Website (About page or main service page)

☐ Google Business Profile (or equivalent)

☐ Primary business directory or listing

☐ One active social bio

☐ One additional reference customers often check →

_____

Step 3: Review for Contradictions (Not Perfection)

As you scan each surface, notice **mismatches**, not missing details. If you find more than 3 issues, **stop**—fixing the top 3 is enough for now.

Ask yourself:

- Do I describe what I do the same way everywhere?

- Does my category stay consistent?

- Would someone be confused explaining this business to someone else?

Common contradictions to watch for:

- Different titles or roles used in different places

- Old services still listed alongside current ones

- Audience changing by platform

- Multiple names for the same offer

Step 4: Choose Up to Three Fixes

(More than three is unnecessary.)

List the **three most impactful changes** that would reduce confusion.

If you notice more issues than this, stop. Clarity improves faster when changes are limited.

Step 5: Make the Changes — Then Pause

Update only what you listed above.

Do **not**:

- Rewrite everything

- Add new platforms

- Change your core description again

Recognition builds when explanations stay steady.

Get the SEEN Clarity Toolkit

You've learned the frameworks. Now put them to work.

The SEEN Clarity Toolkit includes printable worksheets and walkthroughs for everything in this book:

- Topic Trio worksheet

- Customer Question Bank template

- H.A.R.P. Article Builder

- 90-Day Continuity Plan

- "Become Easy to Recognize Everywhere" checklist

- Cue Map walkthrough

These are the same tools from the book—formatted so you can print them, fill them in, and keep your clarity work in one place.

Download the free clarity toolkit at:

seensmallbusiness.com/resource

Appendix

The following sections provide optional context and deeper explanations for readers who want additional detail on specific topics.

This book does not teach tactics or shortcuts under that label. Instead, it focuses on the clarity conditions GEO depends on to work at all.

Research Behind The Order

Optional Context:

(From Chapter 10)

Modern search and AI systems do not evaluate information the way humans do. They rely on consistency, structure, and interpretability to determine whether information can be confidently summarized, compared, and surfaced.

Research from Stanford's Center for Research on Foundation Models (CRFM) shows that language models behave more reliably when information is structured, consistent, and easy to interpret. In their work evaluating instruction-following systems, researchers emphasize that clearly defined inputs reduce ambiguity and make system behavior easier to understand and confirm. Similarly, researchers at MIT studying how language models process information have found that reducing ambiguity in input leads to more stable and predictable outputs.

These findings matter because AI systems do not infer meaning the way humans do. They do not reconcile contradictions intuitively. Instead, they compare patterns across sources and evaluate for repeated, reconcilable signals. When information conflicts, systems hesitate. When information aligns, confidence increases.

Industry research supports this sequence as well. McKinsey & Company, in their 2025 *New Front Door to the Internet* report, describes AI-driven search as a new decision interface that synthesizes information across multiple sources rather than directing users through individual links. In this environment, businesses are surfaced based on how clearly their information can be interpreted and confirmed across reference points. Consistency and clarity make synthesis easier; fragmented or conflicting signals make it less reliable.

This is why optimization works best after recognition is stable. Confirming one clear reference point, reducing contradictions, strengthening existing signals, and then focusing effort where confirmation actually happens aligns directly with how modern AI and decision systems evaluate information.

When the order is respected, optimization becomes quieter and more effective. Effort compounds instead of resetting, and both people and AI systems can confirm what a business does without hesitation.

A Note on Generative Engine Optimization (GEO)

Optional Context:

You may hear the current shift in visibility described as **Generative Engine Optimization (GEO)**.

GEO refers to a change in how discovery works when AI systems generate answers instead of displaying lists of links. In this environment, systems don't simply send people to websites. They interpret information across the web, compare options, and generate summaries and recommendations directly.

This changes what optimization means.

Instead of optimizing for rankings or keywords alone, businesses are increasingly being **described, categorized, and compared** by AI-driven systems designed to reduce choice and simplify decisions for users.

Generative Engine Optimization is the name being used for the work of making a business understandable within those systems — so it can be accurately summarized, confidently categorized, and consistently recommended.

This book does not teach tactics or platform-specific optimization. It focuses on the underlying conditions that make GEO possible: clarity, consistency, and stable meaning across public sources. Those conditions remain valuable regardless of how tools, terminology, or platforms evolve.

Primer: How Large Language Models (LLMs) Actually Work

Optional Context:

> You don't need to understand how AI is built to use it wisely. But having a basic mental model helps you make better decisions about how your business shows up inside AI-driven systems.

This section explains what large language models (LLMs) are—in plain language—and why clarity matters so much.

What Is an LLM and Who Uses Them?

A large language model is a system trained to recognize patterns in language. LLMs power the AI tools your customers use to make decisions:

- AI search tools

- chat assistants

- recommendation systems

- comparison tools

- research and summarization systems

When someone asks, "What's the best option for ___?" or "Who offers ___?" an LLM is often shaping the answer. Large language models themselves are not traditional search engines, but they are increasingly embedded within search platforms and conversational AI systems that influence how businesses are discovered.

This means LLMs are not just tools you use—they are systems customers use to decide.

How LLMs Work (In Simple Terms)

LLMs do not think or reason like humans. They don't understand intent or judge creativity.

What they do is:

- notice which words appear together

- recognize common ways topics are explained

- identify structure in information

- predict what kind of explanation fits a question

In simple terms: An LLM learns **how language is used**, not what it means emotionally.

That is why clear explanations outperform clever ones.

Where LLMs Get Their Understanding

LLMs are trained on patterns across many sources and, when embedded in retrieval systems, may reference multiple sources:

- websites

- listings and directories

- reviews

- repeated explanations across the web

They don't rely on a single post or page. Instead, they check for consistency, alignment, and stable categories.

When your business is explained the same way in multiple places, the system becomes more confident about what you do. When explanations shift frequently, the system must pause and reconcile differences.

Why Clarity Matters More Than Effort

LLMs operate as part of systems, not campaigns.

Systems prefer:

- stable inputs

- predictable structure

- repeatable language

They struggle with:

- mixed messages

- blended topics

- frequent rewrites

- unclear categories

This is why the methods in this book focus on one clear question, one clear explanation, consistent wording, and repetition over time.

You are not trying to "optimize for AI." You are making your business easy to recognize.

The Mental Model That Matters

Here is the simplest way to think about LLMs:

LLMs don't decide who is best. They decide who they can describe clearly.

If a system can understand what you do, place you in a category, and summarize you accurately—then it can recommend you.

If it can't, it won't—regardless of effort.

That is why clarity is not a marketing tactic. It is a system requirement.

Why Small Businesses Can Adapt Faster

While large companies rely on teams, budgets, and complex systems to build clarity, small businesses have a quiet advantage: speed.

With fewer layers and fewer conflicting messages, small businesses can align language faster, stabilize categories sooner, and repeat explanations more consistently.

This book's tools—the Topic Trio, the Question Bank, Content Edges, H.A.R.P., and Continuity—give small businesses access to the same clarity advantages enterprise teams are building at scale.

Across industries, organizations are learning the same lesson: **Clarity is the new competitive advantage.** And in AI-driven discovery, the businesses that are easiest to understand are the ones that get carried forward.

How Large Organizations Are Responding to AI-Driven Discovery

This section is included to give you additional perspective on how discovery systems are changing. It's not required to apply the methods in this book.

Why clarity is becoming a system requirement—not a marketing tactic

Across industries, large organizations are quietly changing how they communicate. This shift is not driven by trends, novelty, or experimentation. It is a response to a structural change in how people discover information, evaluate options, and make decisions. Increasingly, people rely on systems to compare choices, narrow options, and summarize what matters—often before visiting a website, reading a review, or speaking to a person.

In enterprise environments, this is often described using terms like agentic AI: systems that don't just answer questions, but filter, compare, and recommend information upstream of human choice. These systems don't replace judgment, but they shape what information is encountered and how it is framed.

Retailers, universities, consulting firms, financial institutions, and healthcare organizations are all responding to this reality in similar ways. Their goals are consistent across sectors:

- help AI systems interpret their offerings

- reduce confusion in complex information

- support clearer customer decision-making

- preserve the human experience

They are not doing this to appear more technical.

They are doing it to remain understandable in an environment where people increasingly ask instead of search.

This observation aligns with the central idea of this book:

If a message is difficult for systems to interpret, it becomes harder for people to encounter.

When a message is clear, systems can help surface and support it.

The Mental Model Organizations Are Using

Organizations do not need to understand how AI is built in order to respond wisely. But they are updating their communication standards because they've learned something simple:

AI systems reward what they can interpret consistently.

At the center of this shift are large language models (LLMs), which power many tools people interact with every day—including AI search, assistants, recommendation engines, and summarization systems.

Understanding what these systems do—and what they don't—clarifies why consistency, structure, and plain language have become operational priorities, not just stylistic preferences.

Where Language Models Influence Decisions

Language models are often involved when people ask questions such as:

- "What's the best option for ___?"

- "Who offers ___?"

- "What should I choose if ___?"

They are used behind the scenes by systems that compare information, summarize options, and generate explanations.

This means these systems are not only internal tools organizations experiment with. They are part of how customers narrow choices and make decisions—often without realizing it.

What Language Models Do—and Don't—Do

Language models are trained to recognize patterns in how language is used.

They do not:

- think or reason like a human

- understand intent or emotion

- judge creativity or effort

What they do is:

- notice which words appear together

- recognize common ways topics are explained

- identify structure in information

- predict what kind of explanation fits a question

In practice, this creates a simple advantage:

Clear explanations consistently outperform clever ones.

Structure helps systems recognize meaning.

Plain language helps systems describe it accurately.

Why Consistency Matters More Than Effort

These systems build understanding by comparing information across many sources over time, including:

- websites

- listings and directories

- reviews

- repeated explanations across the web

They do not rely on a single post or page. Instead, they review for alignment, repetition, and stability.

When a business is explained the same way in multiple places, systems become more confident about how to summarize it. When explanations vary, systems must resolve uncertainty, which slows recognition and reduces confidence—regardless of how much effort is being applied.

This is why organizations are standardizing language, simplifying descriptions, and reducing internal variation. At scale, clarity is not cosmetic. It is functional.

Why Clarity Compounds Over Time

Language models do not reset their understanding with each new appearance. They accumulate signals gradually as information is revisited, compared, and reinforced.

Over time:

- repeated explanations strengthen recognition

- stable language reduces uncertainty

- summaries become easier to generate accurately

One clear explanation held steady is easier for systems to carry forward than many variations introduced too quickly.

Rather than rewriting endlessly, organizations are stabilizing meaning and letting clarity compound.

The Constraint These Systems Operate Under

Language models don't decide which option is best.

They determine which options they can describe clearly.

If a system can:

- understand what a business does

- place it in a category

- summarize it accurately

Then it can surface.

If it can't, it won't—regardless of effort.

That is why clarity is not just a marketing tactic.

It is a system requirement.

Why This Context Matters for Smaller Businesses

Large organizations use teams and coordination to build clarity across complex systems. Smaller businesses often have a quieter advantage: fewer layers, fewer conflicting messages, and faster alignment.

With consistent language and stable explanations, small businesses can benefit from the same recognition dynamics without enterprise scale.

The methods in this book reflect that reality: fewer ideas, clearly explained, held steady over time. Across industries, organizations are learning the same lesson:

-Clarity is how you stay understandable.

-And in AI-driven discovery, being understandable is what gets carried forward.

Understanding Listings Managers

What Listings Managers Actually Do

A listings manager is a software service that helps distribute and maintain your business information across many third-party platforms at once. In practice, listings managers are also people employed in paid positions whose function is to establish a single "source of truth" for a business's core details—such as name, category, and contact information—and then distribute that information consistently across directories and map services.

They can help identify duplicate or outdated listings and flag inconsistencies that need correction. These types of actions don't determine visibility or recommendation on their own—but they make it easier for search engines and AI platforms to confirm consistency when their systems evaluate whether your business information matches across sources.

Instead of updating your details separately on dozens of sites, a listings manager allows that information to be:

- entered once

- distributed widely

- monitored for mismatches or duplicates

The information typically includes:

- business name

- category

- address or service area

- phone number

- website

- hours

- short business description

Listings managers don't create this information.

They don't decide what your business is.

They don't rewrite your messaging.

They repeat and reinforce whatever information already exists.

Listings Managers Are Part of the Verification Infrastructure

Listings managers sit upstream from AI systems.

They may use automation—and in some cases AI features—to:

- sync data

- flag inconsistencies

- monitor reviews

- suggest responses or improvements

But they are not the systems that decide whether your business is surfaced, summarized, or recommended.

That work happens later—when search engines and AI platforms observe what has been distributed across the web.

Listings managers don't judge.

They amplify.

How Your Business Information Moves Through the Verification Infrastructure

How this system works:

1. You publish your core business information.

2. A listings manager distributes that information as verification infrastructure.

3. Directories and platforms store and repeat it.

4. Search and AI systems pull from those sources when people look you up.

Listings managers sit upstream from search and AI systems. Their role is not to decide whether your business should be trusted or recommended. Their role is to distribute and maintain consistent business information across third-party platforms so those systems can confirm it later.

If your information is clear and aligned, listings managers help reinforce that clarity everywhere it appears. If your information is mixed or outdated, they spread those contradictions just as efficiently.

This is why clarity comes before optimization, and why verification infrastructure supports recognition for businesses—but does not replace it.

Why Most Business Owners Don't Know About Them

Listings managers are not consumer-facing tools. Most business owners encounter listings managers indirectly—because someone else sets up a listings manager platform on their behalf, such as:

- an agency using a listings manager for the business

- a marketing platform bundling listings management into its services

- a franchise requiring a listings management system to keep locations consistent

- a previous vendor setting up a listings manager account years ago

Because listings managers are not well known, they often remain invisible until something goes wrong.

For example: you update your hours on your website, but an old directory still shows the previous hours. Or a past vendor updates one listing but not the others. Now different platforms display different details—and it looks like your business is inconsistent, even though you only changed one thing.

This is the hidden effect of infrastructure: information spreads. When it's consistent, it helps you. When it's mixed, it spreads the confusion just as efficiently.

The system is doing exactly what it was designed to do: repeat information at scale.

Where Listings Managers Fit in This Book

In this book, listings managers are part of verification infrastructure—the behind-the-scenes systems that help business information stay consistent across third-party sources.

That means they can:

- help confirm consistency

- reinforce recognition (being clearly identified as the same business everywhere)

- spread whatever information about your business already exists online

If your business information is clear and aligned, listings managers help that clarity travel farther.

If your information is mixed or outdated, listings managers spread the confusion faster and farther.

Infrastructure supports recognition.

It does not replace clarity.

How a Business Owner Updates Information Through Listings Managers

Most business owners update listings in one of three ways.

1) Update directly on your profiles and platforms (common starting point)

If you manage your own profiles, start with key places like:

- Google Business Profile

- your website

- your main social media bios

- your primary review platform

This is usually enough for very small businesses.

2) Update through an agency or vendor (most common)

If you've worked with an SEO company, marketing provider, or web agency, they may already be using a listings manager for you.

The most important step is simple. Ask:

"Do you use a listings manager to distribute our business information? If so, which one—and who owns the login?"

This matters because you want updates to come from one source of truth, not multiple places that conflict.

3) Use a listings manager yourself (optional, not required)

If you want one place to manage consistency across many directories, you can use a listings manager platform directly.

The workflow is usually:

- enter your official business name, category, and core description

- confirm phone number, website, and hours

- review duplicates or outdated listings

- approve changes so updates can be distributed

Important: a listings manager will spread what you enter—it won't decide what your business should be called or how it should be described.

Clarity still comes first.

You don't have to use a listings manager to benefit from this book—you just need to understand that this infrastructure exists and can amplify both clarity and confusion.

The Understanding Listings Manager Key Takeaway

Listings managers don't decide who you are.

They make sure the internet repeats what you've already decided.

That's why this book focuses on clarity first—before optimization, before tools, before tactics.

Glossary

Ad-Adjacent Answers

When advertisements appear next to or immediately after an AI-generated response, rather than in a separate section. This placement matters because the ad sits close to the interpretation, not just the search results.

AI Visibility

The ability for a business to be seen, understood, and recommended by AI-driven systems. AI Visibility depends on clarity, consistency, and recognizability—not posting frequency or engagement metrics.

AI-Powered Search

A search experience where the system reads multiple sources, synthesizes an answer, and presents it directly — often before the user clicks any link. Used by Google (AI Overviews), OpenAI (ChatGPT), Microsoft (Copilot), and Perplexity AI.

Anchor (or Anchor Page)

The single page or location that clearly explains what your business does, who it is for, and what result it provides. This page serves as the primary reference point other platforms should match.

Answer-First Search

A discovery model where AI generates a direct response to a question rather than returning a list of links. The user receives a summary, comparison, or recommendation as the starting point — not a page of options to browse.

Category (Business Category)

The classification that answers the question: "What kind of business is this?" Categories help AI systems decide when and where a business should appear. Stable categories improve recognition and trust.

Clarity Alignment

The process of organizing and standardizing how a business is described across key public locations so both people and AI systems can recognize it consistently.

Cognitive Load Reduction

The design goal behind AI-powered search: giving users fewer choices and more synthesized answers so decision-making requires less effort. This is why AI interprets before the user engages.

Confirmation

The moment when a person—or an AI system—verifies that a business is legitimate, relevant, and clearly understood. Confirmation typically happens after initial discovery.

Confirmation Focus

The two primary places where a business concentrates its ongoing maintenance so customers and AI systems can easily confirm what the business does.

Decision Layer

The stage where AI systems compare, evaluate, and summarize information from multiple sources to generate a recommendation or answer. This replaces traditional lists of links in AI-driven search. This layer is now also where advertising is beginning to appear, which changes the dynamics of how businesses are surfaced alongside paid placements.

Generative Engine Optimization (GEO)

An evolution of SEO focused on making content clear, structured, and consistent so it can be accurately summarized and cited by AI-generated answers.

Human-First, AI-Readable

Content that is written for people first—clear, specific, and easy to understand—while also being structured in a way AI systems can interpret reliably.

Interpretation Layer

The stage where an AI system reads, compares, and summarizes information from multiple sources before presenting an answer. In AI-powered search, this layer sits between the user's question and the sources — meaning decisions about what to include happen before the user sees anything.

Optimization (as used in this book)

Improving how efficiently a business can be confirmed and matched to the right searches after recognition is stable. Optimization accelerates clarity; it does not create it.

Premature Optimization

Applying optimization tactics before a business is clearly recognizable, often resulting in inconsistent or unpredictable outcomes.

Recognition

How clearly a business can be identified as the same entity across different platforms and references.

Recognition Reset

A focused audit designed to identify and reduce contradictions in how a business is described across key public locations.

Recognition Surfaces

The places people and AI systems check to verify a business's identity and legitimacy—such as websites, directories, listings, review platforms, and brand references.

Repetition (Purposeful Repetition)

The intentional reuse of the same clear message across multiple locations so it becomes recognizable to both people and AI systems.

Source of Truth

The primary location where a business's message is explained fully and accurately. Other platforms should reflect this wording rather than reinterpret it.

Sponsored Result (AI Context)

An advertisement shown within or alongside an AI-generated answer, labeled as "sponsored." Unlike traditional search ads that appear around a list of links, these ads appear near synthesized answers — closer to the system's interpretation of what the user needs.

Structured Input

Information that is organized clearly, consistently, and without ambiguity, making it easier for AI systems to interpret reliably.

Surface

A place where your business information appears publicly online and can be read, compared, or referenced. A surface might be a webpage, listing, profile, review site, or directory entry.

Trust (AI Trust)

The confidence AI systems develop when a business appears consistent, stable, and confirmed across independent sources. AI trust is pattern-based, not emotional.

Trust Signals

Observable indicators—such as consistent descriptions, listings, reviews, and third-party references—that help AI systems confirm a business's legitimacy.

Upstream Interpretation

The process where AI systems evaluate, filter, and summarize information before a human sees it. In traditional search, users made choices from a list. In AI-powered search, the system makes interpretive choices first — meaning visibility depends on being understood before that step, not after.

Verification Infrastructure

The network of directories, listings managers, review platforms, map services, and brand references that repeat and confirm business information across the internet.

Visibility Gap

The disconnect between what a business is trying to communicate and what AI systems can clearly recognize and categorize. Ads within AI answers can widen this gap, because a business that isn't clearly interpretable may be passed over in favor of one that paid for placement nearby.

What Informs This Book

Research, market signals, and observed system behavior

SEEN is informed by research and by what's already happening in the market. It draws from AI research, consumer behavior, and industry analysis to explain how discovery systems interpret and surface businesses. This section shows what shaped the ideas in this book—so you can see the foundation without getting buried in it.

How Research Is Used in This Book

Research in this book serves three primary functions:

1. Explaining system behavior

Academic research from institutions such as Stanford University and the Massachusetts Institute of Technology informs how AI systems respond to structured, consistent, and unambiguous input.

2. Validating observed market patterns

Industry and consumer research from organizations including McKinsey & Company, the National Retail Federation (NRF), and Pew Research Center helps explain why discovery, trust, and decision-making processes have shifted.

3. Grounding guidance in operational systems

Platform and ecosystem research from sources such as Yext, Constant Contact, Semrush, and Meta provides context for how verification infrastructure, cross-source consistency, and citation-based visibility function outside of theory.

About Institutional Citations

Some references in this book cite research institutions or bodies of work rather than individual studies. This is intentional. These references reflect synthesized findings across publicly available institutional research and reports.

Organizations such as the Stanford Institute for Human-Centered Artificial Intelligence (HAI), the Stanford Center for Research on Foundation Models (CRFM), and MIT's Computer Science and Artificial Intelligence Laboratory (CSAIL) publish multiple reports and evaluations that collectively establish principles related to clarity, interpretability, reliability, and structured input. References to these institutions reflect the broader findings of their published work rather than isolated experiments.

This approach is consistent with Chicago Notes & Bibliography standards for synthesizing institutional research.

A Note on Scope

This book is not a technical manual for building AI systems, nor a speculative forecast. It is a practical guide grounded in observed system behavior and translated into clear, usable actions for small businesses and independent professionals.

Where research is cited, it is used to support clarity rather than exhaustiveness. A complete list of references follows.

References

Adobe Digital Insights. *2024 Digital Economy Index*. Adobe, 2024.

Arizona State University. *Public Remarks and Interviews on AI Adoption, Exponential Learning, and Digital Transformation*. Office of the President, Michael M. Crow, 2024..

Blair, Margaret Henderson. "Wear Out: An Empirical Investigation of Advertising Wear-In and Wear-Out." *Journal of Advertising Research* 40 (November 2000): 95–100.

Buffer. *State of Social Media Report*. Buffer, 2024..

Cepeda, Nicholas J., Harold Pashler, Edward Vul, John T. Wixted, and Doug Rohrer. "Distributed Practice in Verbal Recall Tasks: A Review and Quantitative Synthesis." *Psychological Bulletin* 132, no. 3 (2006): 354–380..

Clutch. "Clutch Report: 65% of Consumers Use AI to Research Products Before Making a Purchase." *Business Wire*, January 22, 2026.

Constant Contact. *Small Business Now: The Current State of SMB Marketing*. Waltham, MA: Constant Contact, 2024.

Deloitte. *AI in Customer Experience: Balancing Automation with Human Connection*. Customer experience roundtable insights, 2024..

GWI (GlobalWebIndex). *Connecting the Dots: 2025 Marketing Trends*. GWI, 2025..

HubSpot. *State of Marketing Report*. HubSpot, 2024..

iPullRank. *AI Search & E-Commerce Behavior: How Consumers Browse and Buy in 2024*. iPullRank, 2024.

Lehnert, Kevin. "Advertising Creativity and Repetition: Recall, Wear-Out, and Wear-In Effects." *International Journal of Advertising* 32, no. 2 (2013): 211–231.

Massachusetts Institute of Technology, Computer Science and Artificial Intelligence Laboratory (CSAIL). *Publications and Research on Language Model Reliability and Structured Input*. MIT, 2024–2025..

McKinsey & Company. *The New Front Door to the Internet: Winning in the Age of AI Search*. McKinsey & Company, 2025..

Meta. *Small Business Insights and Trends*. Meta, 2023..

MIT Technology Review Insights. *Trust and the Human Edge in AI-Powered Customer Experience*. MIT Technology Review, 2024..

Modern Retail. "E-Commerce Sites See Low Sales from ChatGPT Traffic, Study Finds—But That Could Change." *Modern Retail*, October 2024.

National Retail Federation (NRF). *Agentic AI, Consumer Behavior, and the Future of Retail*. NRF Research and Surveys, 2024–2025.

Nielsen, Jakob. "10 Usability Heuristics for User Interface Design." Nielsen Norman Group. Last updated April 24, 1994. Accessed January 11, 2026..

Omnisend. "Nearly 60% of Americans Use Gen AI Tools for Online Shopping." PR Newswire, January 2025.

Pew Research Center. *How Americans Use AI Tools in Daily Life*. Pew Research Center, December 2024..

Semrush. *Global Digital Trends and Organic Visibility Report*. Semrush, 2024..

Stanford Center for Research on Foundation Models (CRFM). *Holistic Evaluation of Language Models (HELM)*. Stanford University, 2023–2024..

Stanford Institute for Human-Centered Artificial Intelligence (HAI). *Research Reports and Publications on Language Model Clarity, Structured Input, and Interpretability*. Stanford University, 2022–2024..

Stanford Natural Language Processing Group. *Publications on Prompt Robustness, Structured Reasoning, and Model Interpretability*. Stanford University, 2022–2024..

Thoughtworks. *Human-Centered Retail in the Agentic AI Era*. Thoughtworks Webinar Series, 2025.

Yext. "AI Doesn't Rank, It Cites. And 86% of Its Sources Are Brand-Managed." Yext Blog, October 9, 2025..

Yext. "Why AI Trusts Structure Over Backlinks and Popularity." Yext Blog, December 2025.

Zajonc, Robert B. "Attitudinal Effects of Mere Exposure." *Journal of Personality and Social Psychology* 9, no. 2, pt. 2 (1968): 1–27.